Coyotes I Have Known

Coyotes
I Have Known

John Duncklee

AN AUTHORS GUILD BACKINPRINT.COM EDITION

Coyotes I have Known

AN AUTHORS GUILD BACKINPRINT.COM EDITION
Published by iUniverse, Inc.

For information address:
iUniverse
1663 Liberty Drive
Bloomington, IN 47403
www.iuniverse.com
1-800-Authors (1-800-288-4677)

Originally published by The University of Arizona Press

ISBN: 978-0-595-53243-8

Printed in the United States of America

iUniverse rev. date: 1/20/2010

Again, to my gracious and loving wife, Penny

Contents

Acknowledgments, ix
Introduction, xi
Off to Mexico, 3
Putting My Money Where My Mouth Was, 9
Buying by the Head, 13
Faced with a Real Dilemma, 29
The Crossing That Spelled Disaster, 37
El Gran Coyote y Otros, 49
The Incredible Shrinking Steers, 59
Leave It the Way It Is, 65
Chicaro, 71
Compañeros, 83
Paying the Bills, 107
Cabezón, 111
At the End of a Halter, 115
My Last Horse Race, 119
Carne Asada on the Hoof, 123
The Passing of a Friend, 127
Guessing Weights, 135
In Retrospect, 141
Index, 147

Acknowledgments

I would first like to thank Robert R. Humphrey, my teacher and friend, who has always offered encouragement and enthusiasm. Bob's understanding, and continued search for understanding, of the land and its systems has been my constant inspiration.

Many thanks to Alan Schroder for his unique editorial expertise, and caring, and to the other staff members of the University of Arizona Press for their many efforts in behalf of my books. I would also like to express appreciation to Robin Maly, the Alpine, Arizona, librarian, for her help in finding and acquiring materials from the far ends of the earth. I am grateful to J.P.S. (Joe) Brown for his inspiration, friendship, and help in filling in the gaps. Thanks to Tom Sheridan for his insights and pertinent commentary on the manuscript. Finally, thanks are due to my first editor, Penny, a constant joy, who somehow manages to edit my writing and still design and make beautiful pottery and paint sensitive, beautiful watercolor creations.

In the chapters that follow, I have changed the names of coyotes and others.

TO Yuma

A R I Z O N A

Santa Cruz River

Tucson Mtns.

Sierrita Mtns.

TUCSON

UNITED STATES

MEXICO

Arivaca Jct.

CANOA RANCH

AMADO

ARIVACA

TUBAC

PATAGONIA

SASABE

TUMACACORI

HARSHAW

POZO VERDE

SASABE

SAN LUIS RANCH

NOGALES

NOGALES

Rancho Agua Nueva

Rancho San Vincente

SARIC

Río Altar

LA REFORMA

CANANEA

TUBUTAMA

OQUITOA

ATIL

IMURIS

Río

ALTAR

MAGDALENA

Magdalena

CABORCA

SANTA ANA

PITIQUITO

S O N O R A

CARBÓ

HERMOSILLO

MT

Map by Michael Taylor

Introduction

I n 1959, after surviving the drought of the 1950s on a ranch on the western slope of the Sierrita Mountains southwest of Tucson, I sold my herd of cows. Then I moved my wife at the time (who was involved with quarter horses) and my daughter to the outskirts of Tucson and began looking for something to do that involved cattle. It happened that at that time the livestock market was enjoying a boom and the stock market was experiencing a slump, so I put the majority of my meager working capital into securities listed on the New York Stock Exchange, the largest legal gambling casino in the world.

But I was too young to sit in the boardroom of E. F. Hutton in the Pioneer Hotel and get vertigo watching the ticker tape trot by. I needed to work. During the next few years I became involved with Mexican steer buying, farming, and raising registered cattle and quarter horses. I discovered that there are many wonderful people out there, most of whom are honest and easy to do business with. There are also some coyotes.

The term *coyote,* when used to describe a man's or woman's personality, is not complimentary. It implies that the person has a tendency to be untrustworthy or sly and cunning like the canine counterpart. Nowadays the term *coyote* is commonly used to refer to those who are paid to assist people in illegal entry into the United States from Mexico. The coyotes described in this account were not involved in this illegal activity.

Human coyotes can be great company and great fun to be with, but transacting business with them can be risky unless one has an awareness of their character traits. Certainly not all the people I met and did business with were coyotes; many were superlative human beings. The experience during this period of my life was fun, frustrating, maddening, interesting, sad, enlightening, humorous, and valuable. The world of business seems to teem with coyotes, and business covers a broad spectrum of life. My own naive outlook toward business and people doing business continued after I sold my herd of cows following the drought. Idealism is often difficult to deal with, both outwardly and inwardly. Yet hope seems continual, interrupted only briefly by occasional situations or transactions involving a coyote, who often appears when least expected.

It is a privilege to try to capture these times along the border between Arizona and Sonora because at this writing the region has undergone major changes. The migration from the south to the maquiladoras on the border has overpopulated Nogales, Sonora, and has become a major influence in changing the way of life in Nogales, Arizona, and farther north in the Santa Cruz Valley. Beyond this, Nogales has become the major port of entry for Mexico's west coast produce, and where the smuggling of Scotch into Mexico was once a lucrative enterprise, the major direction of illicit trade has changed. So has the way of life.

The characters—coyotes and others—I met during the early 1960s reflected the times, and their stories lend personality to the region and the time. Rather than disparaging the human coyotes, it is often possible to learn from their cunning ways.

Coyotes I Have Known

Off to Mexico

*F*luency in a foreign language is generally an asset, for a myriad of reasons. But when that ability lures one to a foreign land to do business, one needs to know more than the language. In late 1959 I agreed to enter a partnership that involved buying Mexican steers to import into the United States.

While prodding my cow-calf ranching operation through the drought of the fifties, I had purchased a pen of Mexican steers at the local livestock auction. These fourteen were animals quite different from the Hereford cattle I had been used to. I had bought them in the fall of 1958 after the winter rains had begun to break the drought. It was a gamble, to be sure, but I couldn't resist the chance. The steers averaged four hundred pounds, and I paid fourteen cents a pound for this conglomeration of shapes, sizes, and colors.

The rains continued into early spring. My Mexican steers gained over two hundred pounds per head, and I sold them for twenty-four and a half cents a pound. My gamble had paid handsomely, and after I sold the cow herd I remembered those fourteen steers when an acquaintance of mine approached me with his scheme to make money in the Mexican steer-buying business.

The steer business involves far more risk than a cow-calf operation. Steers don't reproduce, and if the price has fallen when they are ready for market, a loss is possible, depending on how much weight they have gained and the extent of the drop in the price per pound. Trading in steers is even riskier unless the trader already has a buyer waiting for them. Mexican steer trading is the riskiest of all.

The experience with my fourteen head didn't make me an expert in the Mexican steer business by any means. I was to discover just how much of an expert I was not! I had purchased the first pen of steers in the United States. Someone else had gone through all the hassles of buying them in Mexico and crossing them into the United States. At that time I had pasture for the fourteen steers, but after selling out the cow herd I had moved to Tucson. To my dismay, the market didn't move as favorably as when I bought and sold that pen of profit makers.

Based on my earlier success, I decided to accompany my acquaintance to Sásabe for a talk with Toribio Rodríguez, the purported owner of Rancho San Vicente, a few miles north of Saric on the Río Altar in northern Sonora. We needed a ranch to assemble our purchases. While we were there, we were introduced to Rafael Castillo, the local brand inspector, who said that about fifty head of steers were for sale in the *ejido* surrounding Sásabe. The local ejido grazed cattle on a common pasture. Castillo offered to help us find other steers in the vicinity that we might purchase. I asked Toribio why he didn't buy steers himself to put on his ranch. The reason was Toribio didn't have any money. The next step was to have a look at Rancho San Vicente. My future American partner didn't think that was necessary, but I insisted on at least a day-trip to see the kind and condition of the range where we would be collecting and holding our steers. We made the appointment for the following day. Toribio was to meet us in Sásabe at seven o'clock to lead us to his ranch.

The two-track road from Sásabe to Rancho San Vicente proved interesting and scenic. It passed through Rancho Agua Nueva, owned and operated by an American, Lee Fisher, who had married into the Camou family, who were prominent in Sonoran financial and ranching circles. Between Sásabe and

Rancho San Vicente there were sixteen gates to open and close, and I was sitting in the seat by the righthand door. That position automatically meant that I had the duty of opening and closing all the gates we encountered. On the return trip I made sure that I sat in the middle.

According to Toribio, his ranch encompassed fourteen square miles—nine thousand acres, not a large area for a cattle operation. It would, however, serve as a holding pasture to maintain the steers we bought until we had accumulated enough to be worth crossing into the United States. Rancho San Vicente looked capable of grazing two hundred head of steers for a short time but not strong enough to maintain that number year-round.

I was to buy the steers wherever I could and manage the ranch operation in Mexico. Toribio's part included driving the steers to San Vicente, taking care of them in the San Vicente pastures, and when they were sold, driving them to Sásabe to cross the border. My American partner's responsibility was to sell the steers before they crossed, if possible, and arrange for all the Mexican and American permits for Toribio's signature prior to crossing. The steers were to carry Toribio's brand, and any profits were to be split three ways.

The prospect of working in Mexico excited me, and the deal appeared to offer the possibility of making money. I agreed, Toribio agreed, and my partner and I went to the bank and opened our account.

I owned a 1947 Jeep and a 1955 Chevy pickup with a stock rack I had welded together from pipe. The Jeep had come off the assembly line before the disengaging front wheel hubs for four-wheel drive came into being, so its highway speed was limited to thirty-five miles an hour. I welded a tow bar to the front bumper, put the transmission in neutral, and towed it behind the pickup to Sasabe. My plan was to leave the pickup on the U.S. side of the border and use the Jeep on the rough, primitive roads and trails in the Altar Valley of Sonora.

The first barrier I confronted surprised me. One of the Mexican customs officers told me that I needed a Mexican work permit in order to buy cattle, and the closest place to acquire the document was from the Mexican consul in Tucson. I drove the Jeep back through the gate, hooked the tow bar to the Chevy, and drove back to Tucson. I found that in order to apply for a Mexican work permit I needed a U.S. passport. Once I had the Mexican work permit I was told that I had to have a bond on the jeep. Back to Nogales for the bond. A gigantic bureaucratic runaround! It took a month to get all that accomplished. I had learned not to state my business at the border to that particular customs official. He was looking for *mordida,* a bribe, and I hadn't realized it. But I was beginning to learn. Since then, when crossing into Mexico I have always just said I'm a tourist.

There were basically three classes of steers available to purchase in the area: "Improved Mexicans," which were Hereford or primarily Hereford; crossbreds, which were mixtures of two or more breeds such as Brahman and Hereford, Hereford and Shorthorn, and the "Heinz 57" variety; and *"corrientes,"* which came in assorted colors and shapes. Most corriente cattle come from the Sierra Madre ranges and are the closest descendants in Sonora of the cattle introduced by the Spaniards. Each class commanded a different price. The price of the cattle in Sonora reflected the cattle market in the United States. During the period when I was buying steers for export, the prices were by the head as opposed to by the pound or kilo. The by-the-head price kept the buyer on his toes. One had to be able to judge the weight of a steer or risk paying the same for a steer weighing 300 pounds as for a steer weighing 250. The importance of this skill showed up when the steers were eventually sold in the United States, mostly by the pound. A buyer of Mexican steers wouldn't remain in the business long if he didn't have a good eye for weight.

A number of Mexican steer buyers were operating in northern Sonora when I began my adventures. Most dealt in greater volumes of steers than I did because I was limited partly by the San Vicente ranch's holding capacity and partly by my unwillingness to involve all of my working capital in the experiment. I had learned a valuable lesson from raising calves during the drought of the fifties: When there's a drought, nobody wants cattle. On top of the possibility of a drought, I was in a foreign land with different rules and in a culture that contrasted greatly with my own. The greatest cultural contrast in the world is along the U.S.—Mexico international boundary.

There were a number of Mexican government regulations to comply with before we could cross steers into the United States. The most important was the requirement that we obtain a crossing permit, which I discovered could be based on a whim of the Mexican government. Only a Mexican citizen or an *imigrado* (a person going through the first stages of immigrating to Mexico) could obtain a permit to export steers. I held neither status, but I had Toribio to sign for the permits, so I didn't expect any difficulty in that area. In addition to being citizens, however, exporters needed to employ Mexican and U.S. customs brokers to complete the complicated paperwork required by both governments to allow the crossing. The Mexican brokers were also adept at handling the mordida to be paid to the bureaucrats. Supposedly, my American partner would handle everything that involved crossing the steers.

The buyers of large steer herds could simplify the process because generally the animals were under one brand and thus one owner. The buyer could also reduce the risk by putting up a deposit for the steers at the time of the sale and paying the balance once the cattle had arrived at the border-crossing corrals. In many cases the seller wouldn't be paid in full until the steers had been approved by both government inspectors. Some buyers arranged that their deals would not be paid in full until after the steers were on the U.S. side of the border.

I had none of these alternatives because I was buying small bunches of steers from several sellers, assembling them on the Rancho San Vicente pastures, and holding them until I had acquired enough to make a worthwhile crossing. My partner and I had decided to buy two hundred head before attempting to export. In this manner we could buy Herefords and crossbreds, get them across the line, and separate them into groups that had some uniformity, which would make them easier to sell. It seemed like a very workable agreement, and I launched myself into this new business with energy, enthusiasm, and—in spite of former dealings with cunning cattle buyers and a cow-thieving neighbor—naivete.

Putting My Money
Where My Mouth Was

Whenever I had the chance, I stopped by the E. F. Hutton boardroom in Tucson to see how my portfolio was doing. One day a man approached me. I saw him coming but didn't recognize him. He introduced himself to me, and I realized that I had known him years before in New England. I joined him for lunch in the Pioneer Hotel bar next door. We filled in our lives from the last time we had been in New England, and I learned that he was considering buying a farm in the Santa Cruz Valley. He asked me to accompany him to look at the place.

The following day we drove south and looked at the farm. He was also interested in two sections of grazing land across the highway. The price of the farm seemed reasonable to me, but the grazing land—1,280 acres at a hundred dollars an acre—didn't seem like any bargain. I was surprised when he asked me if I wanted to be partners with him on the farm.

"If you think the farm is a good buy," he asked, "will you put your money up to be partners?"

"Hell yes," I replied almost immediately. Then thoughts began to rush through my mind. I had just committed myself to buying an irrigated farm! What the hell do I want with a farm? How am I going to come up with the money? This guy doesn't know a damn thing about farming, so I'm going to have to make all the decisions. Then I thought back to the ranching experience I had just been through during the drought. An irrigated farm was a place I could raise cattle without having to look at the sky for rain clouds every day, and I was already involved with the Mexican steer deal at Rancho San Vicente. We drove back to Tucson.

Two days later second partner and I met with the sellers. No broker was involved. We bought the farm, and he bought the two sections of grazing land. A corporation needed to be formed because I was leery of partnerships. I didn't want to risk a partner accruing debts and leaving me to take responsibility for them. During the course of our discussion I suggested that I would be willing to assign two shares of my fifty shares in the corporation to our attorney. Should something happen to me, my partner would get the two shares and would thereby hold a majority of the stock. My partner felt good about the suggestion and added that he would do the same to put me in a like position if something should happen to him.

The farm had been leased to a local man who had ensilage still left in the pit silo. We agreed to allow him to use our corrals to feed the remaining ensilage to his cattle.

Rudolfo Montez had been working on the place for a number of years. We had talked with him about remaining to work with us after the sale went through. I told my partner that we should increase his salary because he would be a valuable employee. Rudolfo had a reputation as a very knowledgeable farmer who not only knew how to run the machinery but also how to grow crops. Rudolfo, his wife María, and their four children lived in a small house near one of the large barns close to the fields. "Be good to Rudolfo and this operation will succeed," I said to my partner.

The farm had around two hundred acres, about half of it under cultivation. Fifty-three acres could be planted in cotton under a government allotment. The rest could be seeded to most anything else. Growing hay would have suited me fine because I had enjoyed that in Wyoming and Alberta, Canada, where I had worked on ranches during summer vacations in college. Raising hay in southern Arizona involved tractor-driven machinery entirely, however, while I had experienced haying only with teams of horses. I remembered the countless hours and days sitting on the seat of a horse-drawn mowing machine or a sulky rake behind a team of Percheron horses, talking to them, smelling the new-mown hay, building windrows to bunch into piles for the buck rake to gather and push to the haystack. Growing hay in southern Arizona, in contrast, would have involved a substantial investment in machinery and costly irrigation. Summer rains in the valley are risky for growing hay. They are likely to come just in time to ruin mown hay when it's still on the ground. We figured that we could get more income from some other crop. Besides, any hay cut after July 1 could be ruined by the summer rains.

We decided to plant and cultivate our first crop of cotton and Hegari sorghum, a maize plant used for either grain or ensilage. Rudolfo also suggested that we plant Mexican June corn to sell by the ear for *tamales elotes,* green corn tamales. We bought a tractor and the other machinery and left the planting schedule to Rudolfo. He knew from a lifetime's experience of farming in the vicinity the best time to plant the different crops. It was an interesting education to watch and listen to Rudolfo Montez. He should have been a professor at an agricultural college, teaching bilingual agronomy. Rudolfo's presence on the farm made me feel comfortable pursuing the steer buying business in Sonora. I was sometimes gone for days at a time in Mexico.

We commuted from Tucson to the farm. I was living in a rented place on three acres so that I would have a place to keep two quarter-horse mares. My partner lived in a big house in a fancy subdivision. I left Tucson at five o'clock in the morning to arrive at the farm at six. I could have arrived later, but I was accustomed to rising early and getting to work early. A pot of María's delicious hot coffee was always ready when I arrived at the farm at six o'clock in the morning, and Rudolfo and I enjoyed a half hour or so of conversation. My partner generally arrived at the farm around nine.

I generally made the return trips to Tucson after dark. The narrow, two-lane Tucson—Nogales highway was commonly known as the Camino del Muerte, the Road of Death, not only because of its narrowness but also because of the many accidents involving people returning to Tucson after drinking in Nogales, Sonora.

There were other occasional hazards to deal with as well. One night I was particularly late returning to Tucson. Just beyond the Canoa Ranch entrance my headlights picked up something in the road as another vehicle going south came around a curve toward me. I jammed on the brakes, but the obstacle, an old pickup truck from Sonora stalled without lights, was too close. I couldn't pass to the left without colliding head-on with the vehicle coming south. My only escape was to turn into the ditch to the right, off the highway. I was fortunate that the pit was shallow, and I traversed it without flipping the Chevy.

I stopped, turned around, and drove up to the front of the stalled vehicle. I yelled at the two men bent over the hoodless engine to get in and steer as I pushed them off the highway. I was shaking with fear and anger from the incident, and I admonished them in Spanish never again to park on the highway at night with no lights. The Camino del Muerte had nearly lived up to its name.

Buying by the Head

Toribio Rodríguez had six children, and his one desire for them was to be able to send them to school in the United States. Toribio had once worked for the Buenos Aires Ranch near Sasabe on the U.S. side of the border. His job involved picking up any Buenos Aires cattle that had strayed into Sonora. He knew a smattering of English, but I conversed with him in Spanish to be sure he understood. He had a large *pansa,* belly, and walked like he would rather be horseback. His sharp, thin nose separated his blue-tinged green eyes. In spite of his sixty years, he had retained most of his hair, which he kept under a straw hat most of the time except when he entered someone's house. The residents of the entire area from Sásabe to Saric referred to him as Don Toribio out of respect for his age and status as a ranch owner. That status was brought into question by one of Toribio's cousins, but the controversy never affected the agreement my steer-buying partner and I had with the man.

Within two weeks after my first visit, I returned to Sásabe, drove to San Vicente, and spent a week with Toribio buying a few steers. At one ranch the man had three steers for sale, and we walked over to his small brush corral to have a look at them. The steers were fair-quality Herefords. By the length of their horns I thought they should have been larger, but I dismissed the thought, realizing I was in Mexico. Cattle raising practices were different from those I was used to in the United States.

"How much do you want for your steers?" I asked.

"Sixty-five *oro*," he replied promptly. Oro, or gold, meant U.S. dollars as opposed to *plata,* silver, which meant Mexican currency. Sixty-five dollars a head turned out to be the going price at the time, but I offered him sixty because I didn't think the calves were top quality.

"Roberto Zúñiga is paying sixty-five," he said.

"Bueno," I said, "sixty-five."

I learned quickly that the steer market in northern Sonora was based on recent best prices, more than likely those paid by buyers that dealt in far greater volumes of cattle than I could. The way this market news spread was amazing, given that there might have been only one telephone in any particular town and most of the sellers of small numbers of steers lived away from these towns in circumstances similar to my friend with the three steers. I became aware of the going price and never argued with a man who had only a handful of steers to sell.

We walked over to the small, windowless, doorless adobe dwelling and sat down at the rough-hewn table. The man's wife, showing signs of impending childbirth, placed a variety of chipped coffee cups in front of us. She returned to the small cast-iron stove, where a chipped and battered blue-enamel pot of water steamed from the heat of the mesquite-wood fire. The woman brushed a long hank of black hair away from her eyes as she reached for the sooty coffeepot on a shelf, took a cone of fabric shaped like a miniature butterfly net, and placed it in the container. She measured coffee grounds out of a jar into the cone, grabbed the steaming pot and poured the hot water into the cone. When she was satisfied with the brew, she brought the coffeepot to the table and filled the cups.

I spooned a measure of the coarse, brownish sugar into my cup, stirred it into the dark, almost black coffee a few times and sipped it gingerly. Delicious!

I proceeded to write the check for the three steers. I wondered where he would be able to cash it and how he would get there. As I sat finishing my coffee, two hens and a small pig walked calmly through the door. Nobody thought anything about the visitors, so I assumed there was nothing unusual about the close association. Toribio and the man were conversing about the delivery of the steers. I turned my thoughts to the family and their living conditions. The two young, barefooted children playing outside were dressed in dirty, stained clothing that would be past the rag stage in most American households. While I was in the navy during the Korean War, I had witnessed this degree of poverty in Hong Kong, but I didn't expect to see it so close to the border between Mexico and the United States. But they seemed to be happy.

I thought about the purchase of the man's steers. Here I was in the Altar Valley of Sonora buying steers at a price I hoped would hold long enough so that I could eventually sell them for a profit. And here before me sat a man with my check for the three steers in his pocket and with two children clothed in rags and another child due in the near future to be housed in a shack of a house where pigs and chickens wandered in and out at will. Over the next two years I would drink coffee while writing checks in similar situations, in wonder at how these people's existence contrasted with their apparent contentment with their lot.

Had it not been for the revolution and subsequent land-reform laws, they would have been peasants living on some large hacienda. Some might have been well taken care of, others not so well. After the land reform went into effect, the *ejido* system of collective land tenure was instituted, in which large landholdings were divided up. This was an attempt to change a situation in which 10 percent of the population owned 90 percent of the land. The cry "Land for the landless!" echoed throughout Mexico, and the peasant armies under Villa, Zapata, and others forced a change.

The man with the three steers had been allotted enough land to satisfy the land reform principle but not enough to allow him to make a living. Yet he and his family seemed happy as I sat at the old, battered plank table, enjoying their home-roasted coffee. Obviously their independence meant a great deal to them. Not to be in a state of virtual indentured servitude was the alternative they had chosen. The situation might be compared to the Desert Homestead Act in the United States, by which a man could homestead a section, 640 acres, of desert land and eventually own the title to it after proving up by building a dwelling and living on the land for a certain number of years. One section of desert is not enough area to make a living with cattle, and farming was out of the question at that time because deep-well technology had not been developed when the act was passed.

Both cases seem to have failed because the parcels involved were too small for an economic unit. Perhaps they were both cases of too many people for not enough land, or politicians making decisions about land capabilities about which they knew nothing.

One day we were a long drive from San Vicente, and darkness came before we had settled on a deal for four steers. The people were very hospitable. They didn't ask us to join them for dinner, they just served us. After the meal we sat around telling stories. For the most part, I listened. It got to be too late to drive back over the bumpy roads. That was the first of several nights I slept on a dirt kitchen floor in a northern Sonora ranch house.

After a breakfast of eggs, *chorizo* (sausage made with chile and other spices), and delicious refried beans, we started the jeep and bumped along the narrow dirt road toward Altar. After buying some groceries and salt blocks for the steers there, Toribio and I went back to the Jeep and drove to Saric, a small hamlet but a major village in the upper valley. He knew a farmer who might have some cattle for sale. I parked in front of one of the row houses that lined the village plaza, and Toribio knocked on the old wooden door. A white-haired woman appeared, greeted Toribio, and invited us in. She was Sara Soto, wife of Miguel. We sat down at the table in the kitchen, and as we chatted she made a pot of coffee in the same manner as I had witnessed previously. Miguel came in from the *huerta,* the orchard, at the rear of the house, where he had been working. Toribio introduced me to his friend, who became my friend too for the many years until his death.

In his youth Miguel had spent ten years in the Imperial Valley of California, working on a variety of farms. When he returned to Saric he married Sara, who inherited her father's farmland. Miguel worked that land for many years, and when he reached the age of sixty-two he applied for and received his U.S. Social Security benefits. But he continued to farm in Saric.

Without any questions, Sara made a meal for us. I would experience that Soto hospitality many times and would spend the night on a number of occasions. Our relationship became such that, if I passed through Saric without stopping at the Soto house, I would be scolded the next time we were together. Miguel had sold his own cattle but told us about others that he thought might still be for sale.

The Altar Valley of northern Sonora is filled with small farms and ranches, both large and small. It is a valley oasis, like a ribbon of green through desert and grassland. The grassland had become brushland after years of overgrazing after the Spaniards introduced cattle, horses, and other livestock. The livestock ate the fuel that had once fed periodic fires that maintained the grassland. The shrubs then invaded and became well established. The same syndrome occurred north of the border.

From Rancho San Vicente to Saric, the primitive road follows the Altar River, weaving around the riparian vegetation of cottonwood, ash, and sycamore. There is one large ranch, but most of the holdings are small ejidos. Saric was a farming and ranching town back when I was with Toribio Rodríguez. It was the site of a minor Jesuit mission after Fr. Eusebio Francisco Kino came to Christianize the Pima Indians. The Spanish governor, Diego Ortiz Padilla, appointed Luis Oacpicagigua (known as Luis de Saric) Captain General of the Pimas. It later embarrassed the governor when Luis instigated the Pima Revolt of 1751, which created havoc not just in the Altar Valley but north as far as the Santa Catalina Mountains north of Tucson. Since the 1970s Saric has been a center for drug dealing. Large houses now mingle with the quaint old adobe structures that remain. Others have been torn down. A paved road was constructed from Altar. The new road does not follow the old dirt road that I traveled so many times but instead passes west of all the towns and villages.

From Saric south, the road passed through small farms and at one point left the river to traverse a small mesa where volcanic dikes protrude from the desert. La Reforma is the next settlement. It was small, but we could buy gasoline there from a fifty-five-gallon drum. The road then passed by the site of an old water-powered mill once used for grinding grain. The water was diverted from the river. The old waterwheel was lying in the grass near the adobe millhouse, abandoned after a larger mill opened in Altar.

Just beyond the mill site, the road crossed the Altar River again and then climbed a hill to the town of Tubutama, located on a mesa above the valley. Tubutama has a certain charm, along with the history of another rebellion of the Pimas in 1695.

Kino had established the mission of San Pedro y San Pablo de Tubutama. In 1695 the priest there was a man named Januske. As was customary with the Spaniards and Jesuits, an Opata Indian from the vicinity of Arispe, far to the southeast on the Río Sonora, had been brought in as the overseer on the mission farmlands around Tubutama. This practice of bringing in "tame" Indians as overseers caused elemental friction, increased by the difference in language between the overseer and those being bossed.

Father Januske traveled to Tuape, a small village northeast of Tubutama, to visit a friend during Holy Week. He had not been gone long when the Pimas revolted, killing the Opata overseer and intending to hunt down Januske and murder him. However, after talking things over with some of the elders in the town, they headed west to Caborca instead, where they killed a padre and an altar boy, the first martyrs in Pimería Alta. The rebels were soon captured, however, and were told that only those responsible for the deaths would be punished. Men, women, and children were in the group. Before any of the rebels could come forward, the Spanish soldiers opened fire, killing most of the Indians. The place became known as La Matanza, The Slaughter.

In Tubutama the mission building faces the plaza, and the town overlooks Cuauhtémoc Reservoir, an irrigation project on the Altar River. Walking through the streets lined with row houses made me feel that I had traveled a hundred years back in time.

Toribio and I stopped for a cold beer at the Caballo Bayo Bar, across the street from the church. I noticed a small tire-repair shop next to the church. It seemed strange to me that a bar and a tire shop would be located so close to an old mission church with such historical significance. The facade of the church had been restored, and I could see that maintaining the building's beauty cost money. Yet a tire-repair shop made from rusty corrugated tin and scrap lumber was nestled against its outside wall. A bar just across the street from a church could not exist in Arizona.

After a few minutes of pondering the sight as I sipped the cold beer, I turned my thoughts back to steers. Toribio and I chatted with some of the local residents, who were also enjoying the cold beer, and we had more steers to look at for potential purchase.

From Tubutama there are two roads south, eventually joining the main road from Santa Ana to Altar. We were headed for Altar and drove through Átil and Oquitoa, small farming villages located close to the strips of farmland near the river. Oquitoa, like Tubutama, had an abandoned mill, but here the water was still flowing over the large mill wheel.

Altar was larger than any town in the Altar Valley, but nearby Caborca dwarfed it. From Altar we drove east over the future highway. The roadbed had been started several years before, fulfilling some politician's promise. It had never been completed, however, and had become an almost impassable stretch of washboard road. Most travelers preferred driving over the small two-track road, away from the roadbed that would eventually become a paved highway. We reached the ranch called La Sangre about noon. Toribio inquired about steers as we sat down to a meal of frijoles and pork stew. The rancher had already sold his steers to another buyer. It had been a long drive to find out that information. I began to wonder how long it would take to put together two hundred head. By buying small bunches up and down the valley, I had been able to put together only a hundred so far.

We returned to Saric in the late afternoon, and a man was waiting to show us some steers. I looked into an adobe ruin that served as a corral and saw five nice Hereford steers that were just what I had been looking for. He said that they were samples and that altogether he had thirty-five head.

"Fine," I said, "when can you deliver them?"

"In two weeks," he replied. "They will be here in this corral two weeks from today."

We agreed on sixty-five dollars per head, and I wrote him a check for one hundred dollars as a deposit toward delivery. Then Toribio and I returned to San Vicente.

My accommodation at Toribio's ranch was one room of a two-room structure. One of the rooms, the one having a complete corrugated tin roof, was used for the storage of dried, husked ears of corn. The other room, with half a roof, was my sleeping quarters. The first night I spent at Rancho San Vicente was the only night I slept in the room. Actually, I didn't spend all of even that night in the room because during the night I was awakened by something crawling over my bedroll. I grabbed my flashlight and saw three rats. That was enough of that room for me. I moved my bedroll out into an arroyo of soft sand. I didn't worry about flash flooding because there was no sign of rain, and there hadn't been for weeks. The rats remained near their corn, but just as first light eased into the valley I was stirred out of my sleep by something brushing against my face.

I opened my eyes and found myself staring into the eyes of a billy goat. His whiskers had brushed my face. I don't know how far off the ground I lurched, bedroll and all. At the moment of awakening, when one is not totally conscious, one can see things differently than otherwise. For the split second when I was looking up into that billy-goat's eyes, I thought I was seeing God.

The billy and all his followers skedaddled quickly when they saw the bedroll jump. There must have been thirty head of startled goats. I had not been aware that Toribio kept goats and was not too happy that the steers were going to share the range with them. Apparently the goats had satisfied their curiosity about the creature in the arroyo, and they never again awakened me at first light.

The working corrals at San Vicente were made in *estacada* style. In this type of corral, about every four to six feet, two mesquite posts about six to eight inches in diameter are set about a foot apart so that mesquite poles of varying diameters can be laid horizontally in the space between them. The poles are alternated around the structure to weave the fence for sturdiness. Every so often, after a layer has been placed, all the pairs of posts are wired together to insure against spreading. Over the years, as the bottom poles rot and fall apart, the height of the corral can be maintained by adding poles to the top. The estacada corral is strong and can be built from native trees, as opposed to lumber corrals, which use pine or fir lumber at considerable cost to the builder.

Next to the working corral, a small, round estacada corral surrounded the only well on the ranch. It had been hand dug to eighty feet, and the aquifer below seemed strong enough to furnish all the water that was needed by the household and all the livestock. Pumping the well was a full-time job for Toribio's youngest son, Jesús. He had to keep the mule walking around and around the corral to keep the pump jack moving up and down, forcing the water to the surface and into the long water trough. As the mule, harnessed to a long shaft, walked around his circle, the shaft turned a series of gears that activated the pump jack. Jesús, a boy of six, sat on the corral armed with a handful of pebbles. When the mule stopped, Jesús would throw a pebble at the mule's rump and yell "Andele!" (Get up!). Many times young Jesús tired of his job and nodded off, still sitting on the top of the corral. The mule had an uncanny knack of knowing exactly when the pebble tosser had closed his eyes. At that instant the mule stopped walking.

The ranch house overlooked the arroyo from a low mesa. The building was small and by U.S. standards would be spacious enough only for a couple without children. I never understood how the Rodríguez family found room to sleep without encroaching on one another.

Abelardo, at thirteen the second oldest son, did most of the cowboy work under Toribio's supervision. Pancho, the oldest, had emigrated with his wife and their children to Arizona, where he worked on a cotton farm west of Tucson. The two girls, whose names I could never remember, did the household chores, including part of the cooking under the eye of María, Toribio's wife.

A cowboy named José, who was not part of the family, worked with Abelardo. One day in winter José set out on horseback for Nogales to take care of some personal business. The winter temperatures at night hovered close to freezing, so I gave him an old flight jacket I had acquired when I was in the navy. Otherwise he would have ridden for many chilling hours in a worn-out cotton shirt. I didn't notice José at the ranch for the next month and asked Toribio what had happened to his cowboy. I was informed that José and my flight jacket had decided to remain in Nogales.

Toribio and I returned to Saric on the appointed day to buy the steers I had paid the deposit on. I approached the corral with the man who had promised to have thirty-five head of Hereford steers for me to buy. I looked into the adobe enclosure and saw thirty head of small, gangly crossbred steers. The samples he had shown me two weeks before were not even in the bunch.

"Do you still have my check for a hundred dollars?" I asked.

"Yes," the man said, and took the check from his pocket. After taking the check, I tore it into pieces.

"You told me that you had thirty-five head of Hereford steers," I said. "There's not a Hereford here, not even the samples you showed me. These steers look like a bunch of jackrabbits, and they are not worth sixty-five dollars."

The man didn't say much except, "These are good steers."

"They may be good steers," I answered, "but they're not worth anywhere near sixty-five dollars."

"How much will you pay for them?" he asked.

I looked back at the steers. They would do well to weigh 250 pounds. Although they were crossbreds, I decided I would try to buy them, since Hereford steers seemed to be getting scarce in the region.

"I'll give you thirty dollars."

The man shook his head. "Thirty-five is as low as I can go," he said.

"All right, you sold your steers if you deliver them to San Vicente," I said.

He argued that he would have to charge for hauling the steers. I was still not happy with the situation of his misrepresentation so I told him that the thirty-five dollars a head would include hauling to the ranch or he could try to sell his jackrabbits to someone else. I sensed that if I allowed this man to take advantage of me and make me appear the fool to others who might have steers for sale, the project at hand would be a failure. The steers reached San Vicente that afternoon. I wrote another check.

Two days later we went to a ranch near Saric to look at thirty-five head of crossbred steers that had been raised south of the tick zone. To eliminate cattle ticks, which carried tick fever, the Mexican government and cattlemen's associations had first carried on a campaign to free the northern part of Sonora from ticks by dipping the cattle or spraying them with insecticide containing benzine hexachloride. After the tick-free zone had been established, all cattle originating south of Hermosillo had to be dipped before crossing into the northern part of the state. After dipping, the steers had ear tags clamped to them, and a certificate called a *guia sanitaria* was issued to prove that the animals had been officially dipped.

The thirty-five steers we bought had ear tags, and the seller gave Toribio the guia sanitaria along with the *guia de tránsito,* the brand inspection papers. Some of the steers showed the genes of corriente cattle. One was larger than the others. He was a pinto with long horns that looked at least three years old. I bought him with the rest because I had a special purpose for that fellow: I wanted to break him to ride. The people in charge of the quarter-horse show in Tucson had introduced jumping horses as an adjunct activity. I thought it would be an excellent statement to enter the steer in the jumping contest as "El Pinto," and when the announcer called for him I would ride the steer into the arena.

The Jeep had a dead battery, so I had parked it on a hill near the corrals while we negotiated for the steers. After the deal was made, I attempted to start the Jeep by letting it roll downhill for momentum before releasing the clutch. It rolled easily and proceeded at a comfortable speed that I was sure would be fast enough to make the engine turn over. I released the clutch, heard a horrible sound coming from the transmission, and quickly braked the Jeep to a stop at the bottom of the hill. I tried the gearshift and there were no gears! We borrowed two horses and rode back to San Vicente. Two days passed before someone driving to Sásabe came through the ranch and gave me a ride to the border. I arrived at eight o'clock in the evening. My Chevy pickup was not waiting for me because it was the one trip I had made without towing the Jeep behind as far as the border.

I telephoned my steer partner in Tucson from Sásabe to tell him that the Jeep was crippled and I needed a ride. He grumbled a bit, suggesting that I should have someone in Saric fix the transmission. I explained that there were no mechanics in Saric and insisted that he come and get me in spite of the late hour.

Two hours later I was glad to see my partner drive up in his station wagon.

I spent the next week at the farm, helping Rudolfo as much as I could. The steers at San Vicente would be under Toribio's charge. Sometimes it felt as if I had bitten off too large a chunk with the farm and the Mexican steer deal, but I had seen other people carry on several enterprises at one time. I didn't spend much time in the rented house in Tucson because I wanted to keep a close watch on the businesses, especially the Mexican steers.

My partner on the Mexican steers owned a Jeep in addition to the station wagon and another automobile. The solution to getting my Jeep back was to have him drive his Jeep to San Vicente and tow the cripple back to Tucson, or at least to Sasabe, from where I could tow it behind the Chevy. He chose the latter, so a week later we set out, he in his Jeep and me in my Chevy. I parked at the border and accompanied him to the ranch, opening and closing all sixteen gates.

Toribio had towed my crippled Jeep to San Vicente behind two horses. I asked my partner if he wanted to look at the steers I had bought, but either he wasn't interested or he wanted to get back to Tucson. I began to wonder why he didn't take the opportunity to look at the cattle so that he would be in a better position to sell them as per our agreement.

We hooked the cripple behind the other Jeep, and were on our way. After passing through the first gate into the Rancho Agua Nueva the road leaves the valley. A formidable hill leads to a broad mesa. The road up the hill was deeply rutted and had caused me several hours of consternation in the Chevy, which "high-centered" many times on the way up. After that experience I used the Jeep exclusively in Mexico. My partner's Jeep in low-range four-wheel-drive made the ascent without falling into the ruts, and we arrived in Sasabe without mishap.

I towed the Jeep into Tucson and left it with an excellent mechanic, who replaced the ground-up transmission, the clutch, and the dead battery. A week later I returned to Sásabe with the Jeep in tow behind the Chevy. I stopped at Rafael Castillo's house and discovered that Toribio had come in from the ranch and was waiting for me.

He was excited, saying that he had heard about sixty head of Herefords for sale at Pozo Verde, a ranch west of Sásabe on the border. We drove to Pozo Verde and looked at the steers. They were good-looking cattle for the most part, but there were a few that were too small to be worth sixty-five dollars. I bought all but the smaller steers, with the stipulation that they be driven as far as Sásabe a week later. The week gave Toribio time to get his men and horses in from San Vicente.

Everything went well. The Pozo Verde steers arrived, and my count was 180 steers, 20 head short of the number we had originally planned on. When I returned to Sásabe I heard the bad news: The Mexican government had "closed the line" against cattle exportation to the United States. I asked our brand inspector, Rafael Castillo, what it all meant. He explained that there was a beef shortage in Mexico City and the government had decided to force the price of beef lower by not issuing export permits. I asked him how long the situation might persist. "¿Quién sabe?" he answered. *¿Quién sabe?* seemed to be the answer to all questions pertaining to the Mexican government.

I pondered the surprise, first deciding that I would not purchase any more steers, at least until there was a change in government policy. The sudden edict set me to wondering about the carrying capacity of Rancho San Vicente. Our original plan was to assemble steers there, not to hold them indefinitely. The price of steers in Mexico would slide downward as long as exportation to the United States, the major market, was prohibited. Quién sabe how this entire situation would end up.

I remembered when the border closed to cattle exports during the foot-and-mouth disease quarantine from 1946 to 1953. That border closing was mainly the result of the efforts of Texas cattlemen seeking an excuse to ban cattle importation from Mexico. The outbreak of foot-and-mouth disease happened way south of the border. It was never as threatening to the U.S. cattle industry as claimed by the Texans. It was during the quarantine that an enterprising cattleman, Pépe Rebeil, along with other investors, built a beef canning plant in Magdalena and sold canned beef to the Marshall Plan. According to Jim Garrett, who bought cattle for the plant, at the height of its activity the Magdalena plant's thousand workers were processing a thousand head a day. The business was successful enough for Pépe to accumulate several fine ranches in northern Sonora.

Garrett had been approached by the U.S. government to become an appraiser for the foot-and-mouth eradication program. He decided that continuing to be a cattle buyer in Sonora would be in his best interest. As if to confirm the wisdom of Jim's decision, the news came that one of the U.S. appraisers had been stoned to death in a small village outside Mexico City. Before the program was over, three more Americans were killed. The village folk did not understand why people were arriving out of nowhere to kill their livestock.

The Christmas season approached, and I had an idea that I discussed with my partner. I had noticed the many young children in Sásabe. Many of them looked as if they could use some happiness for the holidays, though the Mexicans didn't commercialize Christmas as we Americans do. My steer partner and I agreed that we would purchase all the necessary ingredients for beef tamales, a Mexican Christmas tradition and would ask María Castillo to hire enough women in the village to make three hundred tamales. We planned to fill paper sacks with two tamales a grapefruit, an orange, and a handful of candy. We could provide a special holiday to 150 children. We also agreed that María should not reveal from whom the little gifts came. We bought all the supplies after María agreed to her task, and we hauled it all from Tucson. The Mexican customs fellow didn't bother to check what was in the Jeep, so I didn't have to explain what we were planning.

I didn't return until after the holidays were over. As I stopped at the Mexican customs house the officer on duty said, "The children really enjoyed the tamales, fruit, and candy." I thought to myself that keeping secrets is difficult no matter what the culture.

Faced with a
Real Dilemma

With the border closed to cattle crossings, the question of how to proceed with the steer operation loomed, along with the constant wondering about seasonal rainfall. Without sufficient rains to replenish the San Vicente pastures, the cost of hay and grain would turn our profit into a loss. Would I have to go through another drought? I met with my partner to discuss the options. We decided that keeping the steers and feeding them would be the most prudent way to survive. That would take cash, so we agreed to make deposits to our account—three thousand dollars each. I went to the bank and made my deposit.

Things at the farm were going well. The mesquite trees had sprouted their new leaves to say that there was no longer any danger of a killing frost. The Hegari sorghum and Mexican June corn had been planted. The cotton had to wait until the middle of April because of a government regulation intended to help control the pink bollworm. I didn't understand the theory too well. After all, I was a cowman, not a farmer, in spite of my degree in agriculture. Rudolfo had everything under control and going smoothly.

I had wanted to learn to fly an airplane since the experiences I had had in the navy, but there, as an aviation structural mechanic, I was always in the back seat. I wanted to become a pilot and maybe buy my own aircraft to fly to Sonora to buy steers, so I went to the Hudgins Air Service at Tucson International Airport to inquire about what it would take to learn to fly. I went up for my first lesson the same day. I had a basic knowledge of aircraft and what made them fly, so the instructor let me do the takeoff and we headed south. I enjoyed being at the controls, performing turns, banks, and stalls at the instructor's orders. After an hour he told me to head back and land. He took over the radio contact with the control tower because I had enough to think about anticipating the landing.

Once a week I went for a lesson when I wasn't in Mexico at Rancho San Vicente. I looked forward to my lessons and began to understand the air traffic controller's garbled messages over the radio. I wondered what day I would arrive at the airport and hear the instructor tell me to take the aircraft up for a solo flight.

On one of my trips to San Vicente in the spring of 1961, I saw that the steers were holding up all right, but by the look of the rangeland we were going to need some hay and grain in the near future. I talked to Toribio about our predicament and our decision to forego buying any more cattle until after the permit situation was back to normal and the summer rainfall had arrived, if it ever did. He claimed that the ranch could carry another twenty head easily, but I emphasized that I didn't want to take any more chances than I had already.

I spent a night with the Castillos in Sásabe. They had added a small room onto their old adobe house near the arroyo. I would spend several nights there with their kind hospitality. As I drove up to the corral by the arroyo, the aroma of roasting coffee beans wafted over the heavy evening air. María was outside tending the beans, which she roasted in a frying pan with sugar. When the beans were roasted and had cooled, she ground them in a hand mill. María Castillo made the best coffee I have ever tasted, and her frijoles were excellent.

There was an aura of welcome there, and I became good friends with these gracious people. The friendship has continued all these many years, long after I bought my last Mexican steer. Rafael owned a pickup with the same-sized bed as my Chevy. When I sold my truck several years later, I hauled the stock rack I had built to Sásabe and gave it to Rafael.

I made occasional trips to San Vicente to see how the steers were holding up. Each time I loaded three 100-pound sacks of cottonseed meal in the back of the jeep and managed to get them across the border at Sásabe without being discovered by the *aduana*. I instructed Toribio to feed this supplement to any of the steers that looked thinner than the others.

The border remained closed to steer exportation, and I could see and feel the effect of the government edict on the population of Sásabe, which depended in great part on cattle crossings for a living. Rumors constantly flew that the ban was being lifted, but the corrals and dipping vat remained empty. Six months with a border closed to cattle crossings was hurting the economy of Sásabe—as well as my own cash flow. Meanwhile, I was needed to help with the work at the farm in the Santa Cruz Valley, so I told Toribio that I would return to San Vicente in two weeks after buying and shipping hay and grain to feed the steers.

The farm looked beautiful. The crops were doing well, and Rudolfo was irrigating. My spirits lifted at the sight as I drove into the barnyard. The Mexican steer operation ceased to be bothersome. The sight of the healthy crops balanced the gloom of my steers stuck below the border without a permit to get them out.

When Rudolfo finished irrigating, he started cultivating our crops with a tractor and a cultivator. Machine cultivation would be adequate until the crops grew too high. At that point we hired a crew of cotton choppers, who worked with hoes and who were paid by the number of rows they cleaned. The three major problem weeds were Johnson grass, careless weed and morning glory. Johnson grass sprouts from seed or rhizomes and is almost impossible to eliminate from a field once it becomes established. It can be good grazing for livestock except when it becomes stressed by drought or freezing. When stressed, Johnson grass stems produce prussic acid, or hydrocyanic acid, which is very toxic to cattle and sheep. Careless weed is similar to Johnson grass in this respect. These plants are considered weeds when they grow in cotton fields.

I drove to Magdalena, fifty miles or so south of Nogales, to see what could be done about getting a load of hay and grain to the steers at San Vicente. I went directly to the Purina dealer, Alberto Donnadieu, and found him to be affable and fair with his prices, not only for the hay and grain but also for the trucking. I bought as much hay and grain as the truck could carry, wrote Alberto a check, and drove to the center of town.

It was a hot day, and I was hungry and thirsty. I walked along the sidewalk by the plaza, looking for a restaurant. The smell of stale beer floated out the door of one place. I entered, hoping to find something to eat as well as a cold Pacifico. Sitting alone at the end of the long bar was a friend of mine, Jim Garrett. He looked up from his beer as I walked in and recognized me before my eyes became adjusted to the dark room.

"Hey, Duncklee," he said, "sit down and I'll buy you a beer."

I greeted my friend and we enjoyed the ice-cold Pacifico beer and talked about the trouble the Mexican government was causing us. Jim had been buying Mexican steers for many years and holding them at a ranch called El Llano to the south. He faced the same dilemma I did at San Vicente, but he seemed to take it in stride, having long experience with the idiosyncratic Mexican bureaucracy.

Jim had grown up in Tubac in the Santa Cruz Valley twenty miles north of Nogales. He lived on the Garrett Ranch there and had made a good living dealing in Mexican steers. He had an excellent reputation in the business, and in rodeo arenas he could always hold his own at team roping. At this writing, the local roping arena was staging the Eighty-Third Jim Garrett Birthday team roping event. He was there to watch. Just a couple of years ago he had been there to rope!

Later I found a restaurant and returned to Nogales, where I stopped in to see the customs broker. I wanted to hear the latest rumor about the possibility of getting crossing permits. Ramiro Ortiz sat at his desk. He had no new information and seemed as discouraged as I was. There were several customs brokers working in Nogales, Sonora, but Ramiro had long specialized in exporting cattle to the United States.

It was not long after that visit to Ramiro's office that I learned about a conflict between the two cattlemen's associations in Sonora. I happened on the information just by chance, and it took me a while to digest what was going on.

The northern Sonora cattlemen's association, located in Nogales, was feuding with the state association, located in Hermosillo. I didn't understand all the reasons for the conflict, but I did come to understand that part of the permit problem stemmed from it. It was also brought to my attention that Ramiro was taking the side of the northern group. The man who explained all this complicated Mexican politics to me advised me to seek another customs broker, Juan Alvaro Corella, who was not a party to the conflict. I went to Corella's office to find out if there was some way to acquire a crossing permit for my steers.

While talking with Corella, I almost felt as if I were a petitioner rather than a potential customer. He made a phone call to Hermosillo and chatted amiably with the secretary of the state cattlemen's association. After finishing his conversation, Juan Alvaro expressed some hope that he could accomplish what I needed. I left the office with renewed hope that the steers would soon be in the United States. I telephoned my steer partner to tell him what had happened and to advise him to start looking around for a buyer. I continued to wonder why he wasn't trying to fulfill his end of our bargain by trying to get the permits. I drove directly to San Vicente. It had been two weeks since I had sent the truck full of hay and grain to the ranch from Magdalena.

As I drove down the old dirt road—made longer in time by the sixteen gates I had to open, drive through, and close—I couldn't help feeling optimistic that I would soon get the steers off the scant feed on the San Vicente pastures and across the line. Approaching the corral, I noticed Toribio, Abelardo, and Jesús inside the enclosure, along with some twenty head of horses and the goat herd. I stopped the Jeep and walked over to the gate. The alfalfa hay looked good stacked up beside the estacada fence. I was amazed to see flakes of the hay dispersed around the big corral, with all Toribio's horses and goats chomping away at it. I opened the gate and walked into the corral. Just to the right I saw a pen full of hogs eating the grain I had bought for my steers from newly made troughs. There had been no hogs in that pen before!

I was at a complete loss as to what to do or say. Toribio ambled up and greeted me with a limp handshake.

"How are the steers?" I asked.

"Regular," he replied.

"Are you feeding them hay?"

"No," he said, "there is still plenty of feed out there."

"Let's go have a look at them," I said.

We seated ourselves in the Jeep and drove into the rangeland. I spotted a small bunch of steers under a large mesquite tree and drove over to them. They were definitely losing a lot of weight. I continued around the area and saw several more groups of steers. All showed signs of weight loss and hunger.

"Dammit, Toribio," I finally said. "I didn't buy the goddamn hay for your horses and goats, and the grain for your newly acquired hogs! I bought it for the steers that are out here starving to death while your horses are getting fat!"

When we got back I walked around the corral, trying to calm myself and put the situation in the proper perspective. Toribio probably believed that the feed in the pastures was adequate. I didn't think his feeding his horses and goats was an act of dishonesty, but I did wonder about the pen of hogs enjoying the grain. The bottom line was that the steers were branded with

Toribio's brand. He could sell them and pocket all the money and there would be nothing I could do about it. I walked back where he stood talking to his sons.

"I think we need to give hay to the steers," I said. "Let's go look at them and see what you think." I wanted to give him another chance to recognize that the steers were losing weight.

We climbed into the Jeep and I drove through the pastures, looking at the cattle again.

"They look to me like they're losing a lot of weight," I suggested. "Why don't you have the boys gather them and put them near the corral so we don't have to haul the hay out here?"

Toribio didn't have much to say about my plan.

"If they start starving, there won't be anything to get permits for," I said.

We returned to the corral and he told the boys to saddle up and gather the steers and put them in the small holding trap adjacent to the corral. While Toribio was giving the boys their instructions, I opened the back of the Jeep and threw in two bales of alfalfa. I then drove out into the trap, cut the wires on the bales, and began scattering the hay. I repeated the process until I had enough hay scattered to feed the entire herd of steers.

When the steers arrived, they went directly to the scattered hay. I could see that they were hungry by the way they went about eating the alfalfa. I also knew that if the steers were going to get fed, I was the one who would have to feed them. The big pinto steer had held up well, but he was hungry too. I spent four nights sleeping in the arroyo to be ready to feed the steers each morning. The four days made a big difference in their appearance. Before I left San Vicente for Tucson, I threw out an extra ration. I didn't know if Toribio would honor my request to keep feeding them. I told him I would return in a few days. My main thought, while driving back to Sásabe, was a fervent wish that crossing permits would be available quickly.

With the June heat settling in, Rudolfo was kept busy irrigating the crops on the farm. I stopped by on my way to Nogales to see what was happening with the crossing permit through my newly acquired broker, Juan Alvaro. While visiting with my farm partner, I asked him about leasing his two sections to pasture the steers if and when I was able to get them across the line. My partner on the steers had done nothing in the way of getting the permit or attempting to sell the steers once they were in the United States.

The Crossing That
Spelled Disaster

My farm partner and I agreed on a price per head per month for pasturing the steers if I ever got them across. He also agreed to put the north well in operation to distribute the steers better once they arrived. Thankfully, there was one less problem for me to concern myself with. Now it didn't matter to me if my steer partner sold the steers or not. They would have a place to go.

Juan Alvaro had good news for me. He had managed to get the permit. All I needed to do was tell Toribio to come to Nogales and sign the necessary papers. I told Juan Alvaro I would have the old man in his brokerage office in two days.

I drove back to Tucson and headed for San Vicente. I pulled into the corral area nearly at dusk. The haystack had diminished, and I saw that the steers were still in the holding trap. There were only four horses in the corral and no goats. The hogs were asleep with full stomachs. I had decided to avoid the issue of the grain.

Toribio was happy with my good news. The following morning he packed a few things in a sack, including all the papers connected to the steers. We left for Nogales after breakfast.

For once there seemed to be no problem at the customs broker's office. Toribio slowly and carefully signed the permit applications as instructed by Juan Alvaro. I knew he couldn't read what he was signing. He told me later as we were eating supper that he hoped that he could get his family into the United States so that his children could attend school. Toribio told me about that wish many times. In spite of his corral full of horses and goats eating up the expensive hay I had bought for the steers, I had developed a good friendship with Don Toribio Rodríguez.

Our crossing date was set for two weeks later. June is not a good month to cross cattle, especially when they must be driven to the border on the hoof. June is hot and dry. The drive had to be done carefully and slowly, and planned so that the steer herd would have water along the way. I left all that up to Toribio. He knew the trail and the water holes. I did request that he have some hay available for the steers en route.

A week before Toribio started the steers for the border, I talked to a man who seemed interested in buying them. I decided to try to sell him the steers before they crossed. If I could accomplish that, there would be no need to hold them any longer and incur further expense. The buyer and I started from Tucson early, trailing the Jeep, as usual, behind the Chevy. I had bought a couple of nice sirloin steaks, planning to be as hospitable to my buyer as possible.

When we arrived at San Vicente, I was happy to see that the steers had been fed their morning hay and were relaxing beneath the mesquite trees around the corral. We stopped at the corral, chatted with Toribio for a moment, and walked out into the trap to look at the cattle. After looking them over, the man seemed interested.

"Well, what do you think of them?" I asked.

"Good bunch of steers," he said, "but I can't give you ninety dollars a head. The boss wouldn't go for that."

I was disappointed, not only with his reaction but also with the fact that he was buying for some boss instead of himself, as he had intimated before we began the trip to San Vicente.

"You can tell your boss that it will take ninety to buy them here. Once they are across the line he can add the duties, broker's fees, and all the other expenses to get them across."

"The boss would probably go for seventy-five dollars a head here."

"Hell, I could sell them for seventy-five here without a permit. Tell your boss I already have a home for these devils, and they will be for sale again after they are grassed out."

I drove the man back to his pickup truck in Tucson without stopping to cook the steaks. I put them, still wrapped, into my refrigerator.

I spent the following three days at the farm. The day before the steers were due at the crossing corrals in Sásabe, I drove to the Canoa Ranch, bought a pickup load of alfalfa, and hauled it up the road to the border. I handed the Mexican customs officer a five-dollar bill and explained that my load of hay was for the steers that were en route from San Vicente. He motioned for me to proceed as he pocketed the five, and I drove over the rough, rocky road into the village. I parked in Rafael Castillo's corral and went into the house.

María brought us fresh, hot coffee as we sat down to talk. I asked Rafael if he had heard any reports of Toribio's progress with the steers. He replied that they would arrive by noon the following day but that there had been a problem with the inspection before the steers left San Vicente. What now, I wondered.

Rafael explained that Toribio had removed the metal ear tags from the steers that had originated below the tick line, and the guia sanitaria was not valid without the tags in the steers' ears. The result was that those thirty-five head had been left at San Vicente. My spirits plummeted.

I asked Rafael what I needed to do to get the steers without ear tags eligible to cross. He explained that it would take a lot of backtracking with the guia sanitaria to the steers' point of origin so that the ear tag numbers could be replaced. I wondered how many months that would take. What else could go wrong? I was at a point of discouragement that would only be lifted if all my steers were in the United States. I wondered whatever had possessed Toribio to remove the damned ear tags in the first place.

Rafael and I talked about many things that evening. María filled our plates with beans a couple of times until I couldn't eat any more. I began to put my worries away while conversing with my friends. There wasn't anything I could accomplish by worrying. I would have to straighten things out with patience, an emotion I had developed while surviving the drought of the fifties. I went to my bed later than usual.

María had breakfast ready when I walked into the kitchen. Rafael was sitting at the table with his coffee. We decided it would be best if I waited for Toribio to arrive with the steers before hauling the hay to the crossing corrals. Since there was no water for them at the corrals, they would have to be watered outside of town in the *tinaja,* a place where a spring arose in an otherwise dry arroyo. We drank several cups of coffee during the morning as we waited. Toribio and his trail drivers arrived with the steers at about eleven.

I was happy to see them strung out and heading for the crossing corrals. I brought up the rear with the pickup load of hay and followed them slowly until the point rider, Abelardo, led them off the road on a direct route to their destination. Toribio looked tired and a bit incongruous riding a small bay horse. We waved at each other before I headed the Chevy up the rocky road that would take me to the corrals. The steers and I arrived at the same time. I stopped the truck a hundred yards from the gate and waited until Toribio closed the gate behind them. I felt relieved to see that part of the operation completed as I drove up and parked next to the corral fence.

Toribio and I walked into the corral as Abelardo and Jesús began unloading the hay and scattering it around the corral. The steers eagerly went for the alfalfa as I watched them.

"How many do we have?" I asked.

"One hundred forty-five," Toribio said.

As I looked over the herd, I suddenly realized that the big, long-horned pinto steer was missing. "What happened to El Pinto?"

"We ate him," Toribio replied, with complete nonchalance.

I felt like swearing at him. I had looked forward to having fun with that steer. I didn't reply, but I had to walk away and pretend that I was still looking over the herd. I liked the way all the steers went for the hay. That showed me they were healthy, and the feed would strengthen them more.

Most people who are crossing steers will not feed them the day before because the customs duties are based on weight. The less the cattle weigh, the less the duties. But I felt that feeding the steers after a two-day drive from San Vicente would help give them more strength to endure all the dipping and driving during the heat of June. I didn't care if the duties would be higher. By the following morning they would be wanting water, but they would have to wait until they were in the corrals at Rancho de la Osa in the United States.

I made sure that all the hay had been scattered around the corral before I left the area. The sun beat down in the early afternoon, so I sought somewhere indoors to wait for evening. I drove across the line to the Sasabe store, where I used the pay phone to call my partner. I expected him to be in Sásabe before the steers arrived, but by two o'clock that afternoon he had not shown. I reversed the charges on the call.

"The steers are in the corrals and ready to cross in the morning," I informed him.

"All right," he replied. "I'll be down there in the morning."

"There are still thirty-five head at the ranch that won't cross because Toribio removed their tick-zone ear tags."

"What the hell are we going to do with them at the ranch?"

"Keep them there until the mess is straightened out, I reckon. Do you have anyone coming down here tomorrow to look at the steers?"

"No," he replied. "There doesn't seem to be much of a market for them right now."

"See you in the morning."

I had given up depending on my partner to carry out his part of the deal. His remark about no market sounded as if he hadn't tried to sell the steers. I had thoughts about partnerships that were not positive and decided that I would never again be partners with anyone on a bunch of steers.

I spent the night with Rafael and María. In the morning, after coffee and beans, I left for the corrals to see how the steers looked. Toribio and his sons were waiting beside the corrals. An hour later, Juan Alvaro's man arrived to manage the Mexican paperwork, and soon after that the American customs broker drove up. I greeted everyone and felt glad that they had all arrived. The American veterinarian arrived at about the same time the cattle prodders from the village ambled up to begin the process of inspecting, weighing, and later dipping the steers.

Toribio's sons led their horses into the corral and began easing the steers into the long chute where the veterinarian inspected the animals for ticks and scabies. The prodders, outside the chute, kept the cattle moving along as rapidly as possible. I kept out of the way, observing the various activities.

Once they were inspected by the veterinarian, the steers were let out of the chute and onto the scale. The Mexican customs official balanced the scale and weighed the steers under the eyes of my brokers. Once those operations were completed, the steers were ready to be driven to the La Osa corrals for dipping.

The veterinarian turned one steer back for scabies, but the local Mexican butcher offered to buy the animal from me for ninety dollars *oro*. I signed over the brand inspection paper to him and put the ninety U.S. dollars in my pocket. As the steers started down the short trail to Rancho de la Osa, I settled the accounts with both brokers and drove the Chevy back across the border. When I reached the La Osa corrals, I saw my partner sitting in his station wagon. I waved but didn't greet him otherwise.

The dipping process was the last critical operation I was concerned about, because once in a while an animal can drown in the dipping vat. I opened the gate and walked over to the dipping corral just as they started pushing the first steer into the vat full of water and insecticide. The approach to the vat was by another chute that opened on one end of the twenty-foot-long cement trough. The animals jumped into one end, and as they swam to the other there were men above who pushed their heads under, using long poles with a curved, round steel rod attached. The curved steel was placed behind the animal's head to force it under the surface—total immersion. The far end of the vat was sloped and slatted so that the dipped steers could easily walk out of the vat and into a large corral. I kept a close watch to make sure that none of the steers came close to drowning. I don't know what I would have done besides yell a warning to the men wielding the head-dunking poles if I had seen a steer in trouble.

I was glad that I had brought in the load of hay the day before. The steers seemed strong in spite of the temperature. The June heat was intensified at the corrals, which were near an arroyo but lacked shade of any sort.

I had ordered the cattle trucks for three o'clock in the afternoon to ensure that all the operations were completed before they arrived. By 1:30 all the steers had been dipped and were standing dripping in the corral. My partner was outside the corral, looking over the fence.

I was just about to leave the corral to seek the shade of a large mesquite tree nearby when I saw one of the steers acting very peculiarly. It raised its head in a wavering sort of way, and after a minute began staggering backwards. Then it began a wavering trot in a circle. Before I could call out to anyone, the steer ran directly toward the far corral fence, made of two-by-eight lumber. I thought the steer would veer away from the fence, but it smashed headlong into it, breaking through as if the boards were made of paper. It then headed out into the mesquite thicket toward the arroyo. I was dumbfounded. I looked up to see everyone looking with amazement in the direction of the runaway steer. One of the La Osa cowboys ran over and tried to patch the hole in the fence as best he could with baling wire. I ran to the fence and looked in the direction of the steer's flight.

He was nowhere in sight. The cowboy patching the fence said he would go after the steer on horseback to bring it back to the corral.

I turned to see another steer lifting its head and walking backward. I yelled to my partner. "What in hell is happening to these cattle?"

"Damned if I know," he replied.

"Maybe it's too hot and the dipping got to them," I suggested. "Get a bucket and let's start watering these bastards down. Maybe that will help."

With two buckets we began scooping water from the water trough, scurrying to the steers, and emptying the buckets over their heads. Soon there were four steers staggering backward out of control. Then there were six, then eight. I handed my bucket to one of the La Osa cowboys and ran to the ranch headquarters to telephone Doc Pickrell in Nogales. Luckily he was not out on a call, and he said he would drive out as quickly as he could.

Meanwhile I returned to the corral to help with the water treatment. Four steers lay dead. Two more looked to be near the end when I walked in through the gate. I felt completely helpless in the situation. I hated watching the animals die as I tried in vain to help them. The dying dollars didn't cross my mind until later, and even then I remained more concerned about the steers.

Doc arrived and went right to work trying to diagnose what was happening. He quickly concluded that whatever was in the dip was entering the steers' hides and attacking their nervous systems, causing the aberrant behavior and leading to death through suffocation. We began injecting the affected steers with Sparine tranquilizer into their jugular veins. More and more steers began going into the shock pattern. We used more and more tranquilizer until Doc said we had better have more flown in from Nogales.

I drove out to the Sasabe road after the small Cessna aircraft swooped down over the corrals, buzzing us to announce its arrival. The pilot landed on the road just before I arrived. He handed me the package of tranquilizer and took off in a cloud of dust for his return flight to Nogales. I drove quickly back to the La Osa corrals, delivered the precious medicine, and continued injecting the affected steers.

By nightfall we were all exhausted and hungry. After sunset we noticed that no more steers were going into the crazed pattern of behavior. My partner drove into Sasabe and brought back some bread and lunch meat for supper. I made two sandwiches, devoured them, and spread out my bedroll outside the corrals. I doubt if five minutes passed before I was fast asleep.

The following day, the steers began reacting again by mid-morning. The La Osa cowboys had dragged twenty-eight carcasses to the boneyard a half mile from the corrals the previous day, and a black squadron of buzzards circled above. I hoped that we could save the rest of the steers with the tranquilizer treatment, but they continued to die. I have never felt as helpless in any situation as I did in the La Osa ranch corrals.

We had decided that the insecticide containing the gamma isomer of benzine hexachloride was the culprit. The dip concentration had been doubled either by accident or on purpose. There was no way to prove the latter.

After a week I drove to the farm in the late afternoon. I had to escape the morbid scene for a night. At eight o'clock in the morning, after sleeping in the hay barn, I drove to the Cow Palace restaurant for a cup of coffee. Mike Knagge, a resident of Sasabe, who knew what I had been going through, was standing at the bar.

"Easy Money Duncklee," he said. "I'll buy you a drink!"

"I came in for coffee," I said after laughing at Mike's greeting. "But I reckon a drink might be in order even if it's only eight in the morning."

Ten days after the steers had crossed the border, I telephoned for the trucks again. Fifty-three head were dead in the boneyard. The sky overhead was black with buzzards. We loaded ninety-two steers for the trip to the farm. I drove to Tucson first, and then south. I decided that the flying lessons should be put on the back burner for the present. I was close to my first solo flight, but I never went back.

The steers had been unloaded, and somehow they had escaped from the holding trap. Were these steers doomed to be trouble forever? Rudolfo had saddled a horse and was heading out across the river to bring them back when I arrived. I drove after him to help. The steers had entered a small field and gave us no further problems as we drove them back. "Let's get them under the highway," I said to Rudolfo. "We can locate them tomorrow."

Luckily the steers didn't balk at entering the passageway under the highway, and when they reached the other side they began grazing contentedly. The following day Rudolfo drove half of them to the north well.

The Tucson newspapers had carried the story of the steers dying from the dip in Sasabe. Included with the article was a picture of my steer partner squatting in the corral looking at a dead animal. He was quoted as planning to bring a lawsuit against the government for damages. I had no such ideas because I thought seriously that hiring lawyers would be throwing good money after bad. I was tired of the entire mess.

I received a telephone call from my partner announcing that he had a lawyer lined up to file the lawsuit and that I should go to the attorney's office with the figures. I replied that I was not interested in a lawsuit and without me as a plaintiff there could be no suit. The steers had crossed in my name since I had been the one who had obtained all the permits. I would find another way to make back my losses rather than get mixed up with lawyers.

I went to the bank to look into our account. The statement angered me considerably. When I looked through the deposit list, I noted that my partner had not added the three thousand dollars to match my deposit back when I had purchased the feed in Magdalena. It seemed to me that it was not an oversight on his part. I pondered the situation for a day or two. Having been partners with the man, and having been disappointed with his inaction in fulfilling his part of the responsibilities, I suspected he may have calculated that I would not notice his failure to deposit, since the monthly statements were sent to his address. Two-legged coyotes were not confined to south of the border.

I decided against confronting my partner with my discovery. The account had a balance of less than a hundred dollars. I returned to the bank and applied for a livestock loan to pay the pasture rent.

The steers did well on the two sections of hilly country. I checked them regularly and they showed considerable gains in weight through the summer and into the fall.

El Gran Coyote
y Otros

The house on the farm was small. Rudolfo and María had three children, with another on the way. My farming partner and I discussed the situation and decided to enlarge the place to accommodate the Montezes' growing numbers. The plan included tearing down the wood frame part and replacing it with burnt adobe (adobe bricks low-fired in a kiln). We also planned on adding a large section to the north end. We would remove the wood frame section when María went to the hospital for the birth of her fourth child. In that manner the construction would not disrupt their lives to any great extent.

We hired a man from Tubac to do the adobe work. He was an expert mason, especially with adobe. He carefully lined up everything and went to work on the walls. It was a pleasure to see him work with such care and expertise. The walls went up straight and true. I decided that a roof frame of trusses would be the easiest and strongest. I ordered the lumber, made the pattern, and began cutting and nailing the trusses together to be ready as soon as the walls were finished.

When the plates on top of the walls were securely bolted in place, we placed the first truss against the chimney, located on the south wall. I felt that this would be the strongest method of securing the roof. The first truss was nineteen and a half inches from the south end of the plate. I had planned to add a false rafter to the south to make the overhang, but that would be done after the rest of the roof was in place.

"Put the next truss at twenty-four inches on center to the first," I said as we lifted the truss into position for nailing.

"The first truss is nineteen and a half," my partner said. "Put the rest of the trusses at nineteen and a half."

"If we do that, we'll have to cut all the sheathing boards, and the ceiling tile will have to be lathed," I explained.

"I say nineteen-and-a-half-inch centers," my partner insisted.

We sat on the plate arguing for half an hour until I finally became exasperated with the situation, tossed my hammer to the ground, and climbed down the ladder. "Nail the bastards any way you want to," I said. "I'm going to Tucson."

The following morning I drove into the yard to find all the trusses in place with nineteen-and-a-half-inch centers. I didn't say anything further. After coffee with Rudolfo, we climbed up the ladder and began cutting and nailing the sheathing to the rafters. Later we bought one-by-two boards and lathed the ceiling to have something to nail the ceiling tile to.

I had thought about my partner's stubbornness about the trusses. He later acknowledged that he had been incorrect in his insistence. I had concluded that his reason came from frustration. Originally he had asked me to be his partner because I had experience and knowledge of agriculture, especially cattle raising. I had been making most of the decisions about the farm operation but only after listening to Rudolfo, who had forgotten more about farming cotton and Hegari than I would ever know.

When it came time to nail the second truss, my partner wanted to be able to make the decision even though he didn't know what he was talking about. I decided that I needed to be cautious.

The field nearest the highway was rocky. My partner decided that if we removed the rocks we would have another field to cultivate. More acreage, more crops. He also decided to buy a rock-picking machine he had heard about.

After talking to the manufacturer and reading the brochure about the machine, my partner decided to become the dealer for the Southwest. To become a dealer, he needed to buy five of the machines at dealer's cost. I told him that I wanted no part in the dealership or the machines. I had neither the capital nor the desire to take part in such an escapade. He could justify the dealership as a tax write-off, but I was not in that position.

The machines arrived on a long trailer from some northern state. The man spent several hours with my partner, demonstrating the machine and advising on sales. I watched the demonstration with little interest because it seemed to me that the field would have to be thoroughly ripped before using the rock picker. Rudolfo made the same analysis, and the next day my partner hired a man to rip the field.

The "rock-picker" broke down constantly, but instead of ordering parts to repair the damn thing, my partner, the Southwest's newest dealer, put a new machine on the job.

The cotton, Hegari sorghum, and corn did well. It was satisfying to watch these crops grow under the expert care of Rudolfo Montez. When it came time for the first picking of the cotton, we made a sign to put on the side of the highway: Cotton Pickers Wanted. Rudolfo advised us to have the cotton hand-picked for better grades and cleaner harvesting.

Getting ready to pick cotton included far more than making a sign and placing it along the highway. We needed cotton trailers, a scale, a stairway, and sacks for the pickers to rent by the day. We bought two trailers from a farmer in Marana and a couple dozen long, light canvas sacks from the White House department store in Tucson. Rudolfo and I built the stairway using two-by-six lumber, and my partner bought the scales in Tucson. We had already contracted with the Producer's Cotton Gin in Sahuarita to do the ginning and baling. They supplied us with "picking money." Rudolfo handled that part with his usual honesty and integrity.

The day our sign went up by the highway, a man drove in to offer us a deal. He would arrange for the picking and pay the pickers by the pound. All we had to do was to pay him according to the gin weight of the cotton. It sounded a bit strange, so my partner and I asked Rudolfo what he thought about the man's offer. "The reason it sounds like a good deal to you guys," Rudolfo said, "is he will use scales that weigh short to pay the pickers. He will make his money from the difference in what he pays the pickers and the gin weight of the picked cotton."

We rejected the man's offer. I was amazed that such a practice was prevalent. Cotton pickers went out into the fields and labored bent over under the sun all day. To my way of thinking, they earned every nickel. There was no way I would agree to have someone come in and cheat people.

A bus and several automobiles drove in, loaded with people ready to pick our cotton as soon as the sun had warmed the land enough for the dew to evaporate. Rudolfo was in charge. When each picker had a sack, he told them where to start picking. They went to the field of green-leafed cotton plants with bright white bolls sprinkled on every stalk. I could see the look on Rudolfo's face. He seemed anxious to see the results of his care of the crop.

As the day waned, the pickers came in to the scale with their ladened sacks. Rudolfo weighed each sack and paid the picker by the pound. The picker then went to the trailer, walked up the stairway, and dumped the cotton out of the sack into the trailer. Rudolfo packed the cotton down in the trailer by walking through it occasionally. When it was fully loaded two days later, there was around five thousand pounds of raw cotton to be hauled to the gin.

I hooked the trailer to the Chevy and began the trip north on the highway. I planned to pick up the emptied trailer the following morning on my way back to the farm. I had never towed a cotton trailer before, so I thought fifty-five miles an hour would be fine. Just as I rounded the turn toward Sahuarita I heard a loud noise and looked out the side rearview mirror. It scared me to see the large aircraft tire losing its air and the trailer beginning to fishtail. I didn't use the brakes on the Chevy but let it ride in gear to a stop.

I left the cab and walked back to survey the damage—a blowout, and me without a jack that would lift the weight of the trailer. I released the hitch on the trailer's tongue and drove to the gin. Kenny Smith, the gin manager, gave me a spare tire and wheel, along with a large hydraulic jack with sufficient strength to raise the heavy load. I returned to the trailer parked at the side of the highway and changed the wheel. After throwing the ruined tire into the back of the Chevy, I drove my load to the gin at thirty miles an hour.

The following morning I went to the gin to pick up the empty trailer and return to the farm. As I was hitching the trailer to the Chevy, the head ginner approached me with a scowl on his face. "Your first load of cotton set fire to my gin twice," he said.

"How so?" I asked.

"There were rocks in your cotton."

"How in hell did rocks get in the cotton?"

"You've got some cotton pickers who are trying to make more money. Rocks weigh more than cotton."

"I'll be damned," I said in amazement. Here we turned down the man who was going to cheat the cotton pickers, and now at least two of them were not only cheating us but also setting fire to the gin from the sparks made by rocks hurtling through the metal duct work.

I arrived at the trailer while the pickers were still waiting for Rudolfo to start the day's work when the dew lifted. I walked over to the stairway and climbed to the top, facing the group of people waiting to go into the field. I told them what had occurred at the gin because of the rocks in the cotton and that if there were any more rocks, we would have to change to machine picking. That would mean no work for them. I also told them the same would happen should they be tempted to include green bolls in the cotton to increase their pay weights. "If you doubt the accuracy of our scales," I said, "you can have them tested."

From then on we had no more rocks in the cotton and very few green bolls. The green bolls stain the cotton fiber as it is being ginned, or separated from the seeds. The result is a lower grade of cotton—and thus a lower price for the grower—for the entire bale of five hundred pounds.

I arrived early each morning, and after Rudolfo and I had had our coffee we went to the field to get ready for the day's picking. My partner generally showed up just before the pickers went out into the field. Rudolfo would always greet him with "¿Qué pasó?" My partner had no knowledge of Spanish and would say a simple "Good morning." He had an Hispanic maid in his home in Tucson, so one day he said "¿Qué pasó?" to the woman. She replied, "El tren pasó pero no pitó (The train passed but it didn't whistle)." He asked her to repeat the sentence several times as he attempted to memorize it in order to answer Rudolfo's usual morning question.

Everyone was standing around waiting for the dew to lift when my partner drove up and parked. "¿Qué pasó?" Rudolfo said.

"El tren paso pero no pito," he answered.

Everyone laughed. The partner had no idea why until I told him that he had accented the words incorrectly. *Píto* as he had pronounced with the accent on the *i* instead of on the *o*, is a slang word for penis. His embarrassment was enough to discourage him from any further attempts at speaking Spanish.

When the grass began turning straw-colored, I decided that the time to move the steers was close at hand. The market had still not improved, and my steer partner had made no effort to show or sell the animals. Neither had he taken the time to drive from Tucson to look at them.

I had talked with the man operating the Canoa Ranch feedlot. He told me that he had some steers on a ration of whole cottonseed and ensilage, and his steers were making significant weight gains of nearly three pounds a day. I decided that it was worth holding the steers, not only for the chance of a better market price but also to attain heavier weights.

Rudolfo rounded up the steers the day before the trucks were due to haul them north to the Canoa Ranch. When I walked through them in the corral as the sun began to set behind the Tumacacori Mountains, I was pleased by their appearance. Pasturing them on the grass-covered hills had been worth the expense. They had put on a lot of weight.

We loaded the steers the following morning, and I paid my farm partner the pasture rent. When I arrived at the Canoa Ranch later in the day, I saw that those tough Mexican steers, which had survived the gamma isomer of benzine hexachloride, were content and obviously enjoying their new diet.

One evening as I drove back to Tucson, I had an idea that could be incorporated into the farm and my partner's two sections of pasture. The steers had shown that the grasses were strong. Considering the farm's location next to the highway, I felt that it would make an excellent showplace for registered Brangus cattle, a cross of three-eighths Brahman and five-eighths Angus that was increasing in popularity at the time.

The next day I approached my partner with my idea. He agreed that the farm would be a good showplace and that the two sections of hill pasture would be an inexpensive place to keep a small breeding herd. I did not have the resources to enter the cattle part of the operation, but I agreed to manage it since I was well acquainted with the cattle business. We left for Yuma to look over some two-year-old Brangus heifers raised by Floyd Newcomer, a prominent farmer and Brangus breeder.

Floyd gave us a tour of his farming and cattle operation. The bracero program was still supplying temporary legal farm labor from Mexico. The facilities at Newcomer's impressed me as being far more than adequate in the sense that he went the extra mile to provide excellent housing and dining, two things that were often neglected when illegal workers were used before the bracero program was instituted.

The Brangus heifers looked excellent. We bought twenty-four of the two-year-old heifers and leased one of Floyd's bulls. On the return trip I thought about designing and building pens to feed the young bull calves once they were weaned. With the good-looking black cattle in the field next to the highway, the chances were excellent for us to attract Mexican cattlemen to come in for a better look. I also planned to take the best of the crop to some shows in Arizona and New Mexico. I could even foresee that once the herd became established there would be markets in Texas and California. I was good at dreaming.

A week later the cattle truck with the twenty-four Brangus heifers and the leased Brangus bull arrived at the farm. I wasn't there when it pulled up to the chute and unloaded, but I drove in shortly thereafter. There was a conversation going on about the bull. The truck driver was expounding on what a mean son of a bitch the bull was and warning us that we had better not get near the animal. I listened, but I knew that all of Floyd Newcomer's bulls were broken to halter, so the truck driver's admonitions seemed extreme. "Let's see just how mean a bull we have here," I said, and opened the gate to the corral.

They had put the bull by himself in one corral. I walked in, closed the gate behind me, and slowly started approaching him, talking to him. The bull turned to face me without moving away. I continued my slow advance in spite of his occasional snort as he shook his head and pawed the ground with a forefoot.

"Goddammit, he's going to charge you!" my partner yelled as I walked to within six feet of my black Brangus friend.

"Hey, old man, you're not going to charge me, are you," I said in the same low tone I had been using. I hoped the partner would stop yelling. I didn't want the bull to get scared and shy away from me or decide to charge me after all.

I continued toward the bull until his nose was within reach. Very slowly I extended my hand toward him as I soothed him with my one-sided conversation. In case the bull had decided to give me a chase, I was ready to dodge, but somehow I knew I had his confidence. I put my fingers just above his nose and began to rub him gently as I kept talking. Then I eased my palm up his face until I was scratching him between his eyes, then over the poll between his ears. I moved slowly around and stroked his neck. My last gesture was to put one arm around his neck and pat him. "You're not such a mean bastard after all, are you bully boy?"

I walked back to the audience sitting on the corral fence. The truck driver still held his electric cattle prod. "If you had left that damn thing in the cab of your truck, you might not have riled that bull," I said as I opened the gate to the corral and walked out.

One morning while we were enjoying María's coffee, Rudolfo had left to tend to the irrigating and I was about to leave to take a look at the heifers.

"Wait a minute," my partner said. "I want to talk something over with you."

I had no idea what he wanted to say. I thought it must be something about the operation of the farm, but it was more than that.

"Why don't we change the corporation into a partnership for tax purposes," he suggested.

The word partnership struck a sour note with me. I thought for a minute about how to say no. I didn't want any part of a partnership in which either would be responsible for the other's debts.

"I'd just as soon keep the corporation," I said, thinking that would end the conversation.

"Then why don't you buy me out?"

My partner knew damn well I didn't have the resources to buy him out, and I wouldn't have bought him out anyway. I certainly didn't want a fleet of worthless rock-pickers.

"I think I know what you're driving at," I said. "I don't have the money to buy you out, but I think you would like to buy me out."

My farm partner changed his attitude, the tone of his voice, and the look on his face. I doubt if he was much of a poker player. I saw that I had hit on exactly what he wanted.

For the next half hour he did most of the talking. He wanted to do the right thing by me, etc., etc. I had $25,000 invested, plus a year of time and decision making. I was surprised when he offered me $75,000 for my stock in the corporation. He would retain the cotton, Hegari sorghum, and corn receipts. He would also pay the taxes and mortgage payment. All I had to do was sign over my stock to him. I jotted down the figures on a torn corner from a grocery sack so that I could see it in writing. I agreed to closing the transaction on January 2.

We also agreed that I would become the manager of the cattle operation and continue to work. He said he would build a house for me on the farm. My duties included promoting the Brangus as well as fitting them to show. I would also arrange for permanent pasture to be planted, a small area at a time. The two sections across the highway would carry the nucleus herd of heifers, but I felt that when the young bulls came along they would show well against the green of permanent pasture. I agreed to do this for five hundred dollars a month.

The Incredible
Shrinking Steers

The steers at the Canoa Ranch feedlot were doing well, gaining three pounds per head per day. I looked at them at least once a week to try to sense when their gains became less efficient. The market had improved, but there were few buyers looking for cattle. I went to the livestock auction in Tucson for three successive Saturdays to try to find a buyer. On the third Saturday I noticed Jason Roland, whom I had known for a number of years. He was an order buyer, and he was bidding on crossbred steers that compared to those I had at the Canoa.

I followed Roland into the coffee shop when the sale ended and told him about the steers I had for sale. He said he had seen them shortly after their arrival because he and the Canoa manager had been partners on some other cattle and that he would drive to the Canoa the following Monday to look at them.

Jason showed up at the feedlot at close to noon the next Monday. He walked through the steers as they were eating the cottonseed and ensilage.

"I'll give you twenty-six cents a pound for these Mexicans," he said.

"That's below market, Jason," I replied. "Hell, I saw you bidding four cents higher than that at the sale Saturday."

"Those were better steers than these Mexicans," he said and turned away to leave the corral.

"You can buy these steers for twenty-nine cents, Jason," I said.

"I'll give you twenty-eight and a half, overnight stand without feed or water, ten percent cut, and we'll weigh 'em at seven in the morning next Monday." That meant that the steers' stomachs would be empty of feed and water from the overnight stand and that he could cut 10 percent out of the bunch. These could be any steers that he didn't want for any reason. Generally it would be the heaviest that would be in the 10 percent cut.

"They're your steers," I said. "Write me out a deposit."

Jason wrote a deposit check for $1,500 and handed it to me. I was relieved to know that, after we weighed them on Monday, the steers that had caused me so much grief would belong to someone else. In spite of getting a better price than I would have sold them for back when the steers crossed the line, I would still lose money because of the fifty-three dead ones in the La Osa boneyard. But I did manage to recoup some of the loss by holding them on grass and then in the Canoa Ranch feedlot.

I telephoned my partner to tell him the good news that the steers were sold. He didn't seem impressed, especially when I told him that I planned to pay off the bank loan before sending him his half of the proceeds. I also had another feed bill to pay. I then told Bill, the feedlot man, about the sale, and he said he would have the steers in the working corrals by six o'clock Monday morning.

I arrived at the Canoa Ranch at 6:30. I always try to be early and so have spent many hours of my life waiting for people to meet me at an appointed time. The steers were standing in the corral. I didn't expect Roland to show up before the appointed time because the longer the steers stood in the corral, the more they would shrink. It was 8:30 when Jason drove up and parked. The brand inspector had already arrived, but he was used to waiting for Jason Roland.

The steers looked full to me, not like cattle that had spent a night without feed or water. Jason thought so too and went into the corral. He started chasing them around as if he was looking them over, but I knew what he was doing.

"All right, Jason," I said. "Get up on the fence and tell me which steers you want cut out for your ten percent, and I'll do the cutting. These cattle are mine until they are weighed!"

Roland was somewhat surprised at my remarks, but he walked to the fence, climbed up, and sat on the top layer of mesquite poles of the estacada corral. As he pointed to the steers he wanted to reject, I cut them out and chased them into an adjoining corral. He rejected ten head, choosing the heaviest of the steers. Then we put them on the scales and weighed them. I was pleased with their weights.

The brand inspector took me aside when the weighing was finished. "I'll give you thirty cents for the rejects," he said.

"You bought them, my friend."

Roland and I figured out the amount he owed, and he wrote a bank draft for the steers. He complained again that the steers looked too full, but there wasn't anything I was going to do about it. I agreed to help load them when the train arrived that night.

We weighed the rejects, and the brand inspector wrote me a check. I was very happy and drove into Tucson to cash the check and the bank draft. I told the Canoa Ranch manager to figure out my feed bill so I could pay it on my return.

I drove back to the Canoa Ranch after supper. The train was due sometime that evening. I paid the feed bill.

"Didn't those steers look full to you this morning?" Bill asked.

"They sure did," I replied. "Roland thought so to."

"Well, I have to tell you," he continued. "That goddamn Roland screwed me over on a cattle deal a while ago. So I decided to get even with the bastard with your steers."

"I don't quite understand," I said.

"Well, I went out to the feedlot at three o'clock in the morning, opened the water, and filled the feed troughs full of ensilage."

"For Crissakes," I said. "No wonder their bellies were as full as ticks."

The man's revenge became my advantage. I saw Jason Roland a month later.

"Jesus, John," he said, "those goddamn Mexican steers of

yours shrank nineteen percent from here to California!"

"Aw hell," I said, and went about my business.

The train didn't arrive at the Canoa railroad corral until midnight. We loaded them after I gave the paperwork to the trainman. I drove to the farm and slept in the barn. The following day I drove to my steer partner's house in Tucson and settled the account. He didn't mention the $3,000 he had not deposited. I didn't mention it either.

The following week I made a trip to San Vicente to see Toribio and the thirty-five steers that lacked a guia sanitaria. The steers looked fine from grazing the grass and browse produced by a good summer rainy season. Toribio informed me that the eartag matter had been finally cleared up the week preceding my visit. That meant the steers could be exported, which was good news if I could find someone with a permit who would take my steers across with theirs. Thirty-five head would cost too much on a per head basis since the customs brokers charged a flat fee for traveling to Sásabe from Nogales. Toribio told me that Pete Rebeil had a permit and that I might talk to him about crossing my steers with his.

I went to Pete's home in Nogales. He was very hospitable and agreed to help me get the cattle out of Mexico. His crossing date was two weeks away. Two days later I returned to San Vicente and told Toribio the good news. He assured me that he would drive the steers to Sásabe for the crossing.

Since the disaster with the first bunch, Jim Garrett had lost a good number from the same dip mixture affecting a herd of his steers crossing in Nogales. Jim had influenced the bureaucrats in Washington to change the contents of the dip, so I wasn't worried about losing any more from that cause.

I went to Sásabe the day before the crossing to make sure the steers had arrived. As I crossed into Mexico, I overheard that I had been given a Mexican nickname, El Becero Muerto, The Dead Calf.

I was relieved to see that Toribio had the steers safely enclosed in the corral. I walked through them and noticed that they were larger but couldn't compare with those I had put on the grassy hill pasture and Canoa Ranch feedlot. I had no intention of repeating the process with this bunch, however. All I wanted was to get them out of Mexico and through the auction ring. I stayed with the Castillos for the night, and in the morning my remnant steers crossed the border to the La Osa Ranch corrals for dipping. The truck arrived. I helped load them and returned to Tucson.

The market seemed good the following Saturday, and there was a flurry of bidding on my Mexican steers. They brought an average of $108 per head. After the sale I went to the auction office to get paid. I was pleased to see that the amount was $3,800. I had entirely recouped the $3,000 that my partner didn't deposit and had some left over for the aggravation he had caused me by not doing his part according to our agreement. I wanted to get some money to Toribio, however. There had certainly been no profit for him to share. My steer-buying partner and I had lost several thousand dollars, but I couldn't see leaving Toribio without anything for his efforts. I drove to Sásabe the following week and luckily found him in town instead of at the ranch. Toribio appreciated the money I gave him. He wanted to know when my partner and I wanted to buy more steers. I told him that I no longer had that partner and doubted very much if I would try my luck with Mexican steers again. I saw Toribio a few times after that day, but I never saw the Mexican-steer-buying partner again.

I continued to work at the farm. Except for the time it took to deal with my Mexican steers, I commuted to the farm from Tucson every day. I assumed that, according to the agreement my farming partner and I had made, I would be put on his payroll after January 2.

I checked the Brangus heifers every day and became well acquainted with them. In spite of their uniformity, I could distinguish them as individuals. That ability is part of being a cowman. I noticed that the bull had serviced some of the heifers twice and brought this fact to my partner's attention. I pointed out that there was always the possibility that the bull was sterile. When I saw that it was happening a third time, I telephoned Floyd Newcomer in Yuma. He sent another bull that he had tested and took the sterile fellow back to Yuma.

Leave It the
Way It Is

bout a week before Christmas, my partner invited me to his New Year's Eve party. I accepted even though I have always tended to stay home on New Year's. I arrived at eight o'clock. There were two couples I wasn't acquainted with, and I was not surprised to see our attorney and his wife in attendance. There were trays of snacks, and my partner poured a drink for me. The party went along nicely. At midnight everyone wished each other well for the new year. The two couples I had not known before left shortly after midnight, leaving my partner, our lawyer, and me to stand around in the kitchen while the women sat chatting in the living room.

After several drinks during the course of the next hour, my partner became serious.

"What would you think about changing our deal on the farm?" he asked.

"What do you mean?" I responded. He had taken me by surprise.

"Well, suppose I sell the farm for more than we figured it is worth, and you share the profit. And if I sell the farm for less than we figured it is worth, you share the loss."

I didn't like the idea of changing anything about the deal. I was perfectly satisfied with the transaction, which was due to be consummated the following morning. I also noticed that our attorney was listening intently to our conversation. He had been my partner's attorney prior to our purchase of the farm. "Let's just leave it like it is," I said.

The conversation had made me feel uncomfortable. I decided to leave as soon as possible after changing the subject to the future of the Brangus herd.

The following morning I met my partner in the lawyer's office. I signed over my stock in the corporation to him, and he gave me his check. I went directly to the bank to pay off the $25,000 loan.

I went to work the next day. Working for wages seemed fine to me. I had always been loyal to my partner and continued to look out for his interests as far as the farm operation was concerned. To me, our transaction didn't affect our friendship. I checked on the Brangus and went out to the large hay barn where I had planned to build pens for the bull calves. My ex-partner drove into the barnyard around midmorning. He seemed slightly distant in our conversation, but I had seen him that way before.

Two weeks later I heard that a horse ranch bordering the Santa Cruz River outside Nogales was for sale. I drove out with the realtor to look at the place and thought that it would be nearly perfect for raising quarter horses. The asking price was $90,000. I thought that quite high priced, but I also discovered that the seller was undergoing some financial difficulty from his divorce. I made an offer of $50,000. The realtor said she didn't think the seller would agree to such an offer, but even she was surprised when the man made a counter offer for $55,000. I would have to put up $25,000 in cash and assume existing first and second mortgages.

I accepted the seller's terms and bought the San Luis Ranch. I felt overjoyed with the purchase, and within two weeks the deal closed. I returned to the same bank for another $25,000 loan.

I told my ex-partner that he would not have to go to the expense of building a house for me on the farm. I planned to commute the easy twenty miles from the San Luis. He questioned whether I could manage both operations. I assured him that the horse operation would not interfere with my responsibilities at his farm. He seemed to accept my explanations.

The San Luis Ranch covered 120 acres bisected by North River Road, which turns north off the Patagonia highway. The Comorro Wash ran close to the southern boundary of about fifty acres of rangeland. The Santa Cruz River made a turn around the irrigated fields and formed the west boundary. North of the barn area, an eighty-acre lease came with the ranch, and part of this parcel was tillable.

The previous owner had constructed several corrals made from pipe and V-mesh fencing. These enclosures lined the north side of the entry driveway, and three more had been placed north of the old adobe barn. There was a large paddock of the same construction between the main house and the barn. Next to the barn he had built three concrete-block stalls. A two-acre pasture occupied the area between the pens along the driveway and the barn.

The main house, guest house, and barn were located on a low mesa overlooking the floodplain of the Santa Cruz River. The previous owner had built a swimming pool at the rear of the main house. Next to North River Road, a small dwelling was located close to the entrance to the headquarters. The irrigation well was next to the road, and the domestic well was under a shed to the rear of the barn. The barn was not large but would hold more than a semitrailer load of hay and still have room for the shop and grain room. The entire setup had potential in my estimation, but I would find that there would be much work to do before I could accomplish what I wanted for the operation.

Moving from Tucson to the San Luis Ranch was an interesting experience. My two mares traveled easily in the two-horse trailer. I bought an ancient two-ton Ford truck for $250 to accomplish the rest of the move, which included household stuff and almost a ton of hay. Once I had moved onto the new ranch, I decided to keep the truck, which proved to be a good decision. Shortly after settling in, I read an advertisement in the Tucson paper offering registered quarter horse mares for sale. Early the next Saturday morning, I drove out to a ranch at the base of the Rincon Mountains to see what kind of mares the man had for sale.

The seller was a retired medical doctor living with an attractive young woman whom he introduced as his nurse. He showed me the mares, all of which had foals. I looked at their registration papers to see if the descriptions coincided with the animals. One of the mares varied slightly from the papers, but I did not quibble about it. I felt that I could straighten out that matter in time. He was asking $400 a pair. Three of the mares had fillies by their sides. The other mare had a bay stud colt. I didn't argue with the man, and I gave him a deposit. I told him that I would return for the mares the following day. I hoped that he would have the brand inspector come out beforehand.

The old Ford became important again. I drove it to Tucson to pick up the mares and foals. The truck had been used for furniture hauling, so the box was enclosed. The rear panels were missing, but I managed to fashion an enclosure system that would contain the animals. I planned to halter the mares and tie them, but I was disappointed that one of the mares had never been halter broken. I was glad that the ranch had a loading chute.

Unloading them at the San Luis presented a problem since there was no loading chute. I off-loaded the haltered mares by backing the truck against a large mound of dirt and leading them out. I hoped Mere Wolf, the unbroken mare, and her palomino filly would follow the rest into a pen. All went well until Mere Wolf decided that she had no intention of going into the pen. I took a bucket of grain that I had brought for such a situation and shook it near her. The mare knew what grain was and eagerly put her nose in the bucket as I walked her through the gate and into the pen. I had an instant band of broodmares but no stud.

I spent the remainder of the day with the foals, finding out whether they had been halter broken. I had my work cut out for me because none of the foals had ever seen a halter.

On Monday morning I drove to the farm. I had been working

for my ex-partner for six weeks, but I had not yet received a paycheck. He drove into the barnyard at ten o'clock. I saw him arrive and decided to have a talk with him.

"When is pay day on this outfit?" I asked.

"Hell, John, I can't afford to pay you to work here."

"I wish you had told me this before," I said. "I reckon I'd best be down the road. I need to find something to do."

I had planned on the income from managing the farm. I didn't want to use up my capital for living expenses, and the horse operation was far from paying its own way, much less making money for my living expenses. I told him that if he needed any advice or anything else, to feel free to call me. I was disappointed but not angry.

I drove back to the San Luis Ranch, contemplating the dilemma. I had thought the farm would provide steady work with steady income. I thought about the work I had done in the six weeks while I thought I was on the payroll. It became obvious that my ex-partner didn't want me or my advice at the farm. I reckoned that he wanted to make all the decisions himself. Good luck!

Thirteen years later I discovered what may have been the reason for his reluctance to continue our agreement. The New Year's party scene in his kitchen vaulted back into my memory when a real estate man asked me an interesting question.

"What ever happened to that deal your partner on the farm had going on his two sections of land across the highway?" he asked.

"I don't know what deal you're talking about," I replied.

"Hell, he had a buyer for the two sections for something like three hundred an acre, three times what he paid, and he was going to throw in the farm for nothing to sweeten the deal."

I thought, so that's why my ex-partner wanted to change our agreement on the sale. If I had agreed to share the loss or profit when he sold the farm, I would have lost everything, including the twenty-five grand I had borrowed to begin with.

"I never heard about that deal until you just told me," I said.

"I have often wondered why that deal didn't fly," Mike said.

The conversation ended. I was astonished at what I had just heard. It angered me to think my farming partner would even think of putting me in such a position, much less that he had already made an offer before we talked that New Year's morning in the presence of our attorney. I continue, after more than twenty years, to wonder about that one.

Chicaro

went to work with the foals. The first step was getting them used to the halter. Breaking them to lead took more persuasion and teaching. Mere Wolf's palomino filly was the largest and strongest. The others were easy to catch, but I had to rope the palomino and snub her up to get the halter on. She had a lot of her mother's personality: she fought every step of the way. Breaking her to lead took the better part of a day.

A pretty sorrel mare had a small sorrel filly by her side. The filly was easy to train. A small chestnut mare named Mrs. Thomison II was the mother of the little stud colt, and another palomino mare had a palomino filly. After the foals were all halter broken and had learned to follow the lead rope, I decided to sell them as soon as possible for cash flow.

I telephoned a man in California and told him about the sorrel mare and filly. He liked the bloodline and bought her for $1,250. He agreed to pay me a hundred dollars to deliver the pair to Escondido. I sold the stud colt for $300 to a Mexican rancher I had known for several years. The two Palomino fillies went to a man in Rockford, Illinois. He planned to come for them himself. The largest filly sold for $600, and the smaller filly brought $500. Within two weeks I had sold all the foals and one of the mares. I felt I had made a good horse-trade. Four mares and foals had cost $400 a pair, or $1,600 to begin with. I had sold

$2,650 worth of horseflesh and still had three mares. I realized that I really didn't need the job I thought I had had with my ex-partner. But there were mortgages to pay, and I needed to find something else to make a living on besides horse trading. But before I went looking I received a telephone call from California that was to begin a happy relationship with a quarter horse stud named Parker's Chicaro.

The telephone call came from the owners of the horse. I had visited with these men earlier in Phoenix during the Arizona State Fair Quarter Horse Show, where I had proposed to lease the twenty-one-year-old horse. They had called to say that they would agree to lease him to me and would also haul him to the San Luis Ranch. I couldn't have been happier to hear that good news.

Chicaro had been brought to Arizona from Texas by Wirt D. "Dink" Parker, a cowman and horse breeder in Patagonia. Dink knew horses as well as any man in the country and was successful not only as a breeder but also with his horses on the racetrack. Chicaro did well on the track in Tucson, and when he finished that career, Dink took him to his Salero Ranch in the hills north of Patagonia and put him on the range with a band of broodmares. I had not only seen Chicaro running races, I had seen him with his mares at the Salero Ranch before Dink sold him.

Chicaro's bloodlines were impressive. His sire, Chicaro Bill, was by Chicaro, the Thoroughbred, who was by Chicle, by Spearmint, by Man o' War. Chicaro's dam was Beula Burns by Black Joe, by Little Joe, by Traveler, the foundation sire of the quarter horse breed.

As a sire, Chicaro was best known for the broodmares he produced. My wife at the time owned one of these mares, named Salero Maiden, a grand champion mare at halter. Salero Maiden was one reason to lease the old stud. I didn't realize during that telephone call that Chicaro and I would become close friends.

The appearance of the old stud shocked me as he jumped off the back of the two-ton cattle truck onto the mound of dirt where earlier I had unloaded the mares. His black coat was dull, his hooves were so long and ill-kept that he couldn't walk well, his eyes showed little spirit, and I could see his ribs. I led him down the road past the pens holding the mares. When he saw them, his interest was piqued and he nickered. "Well, old man," I said to him. "I see you're still interested in the girls, so we'll see what some proper care will do for you."

From that day Chicaro would have the best hay I could buy and two coffee cans of grain morning and evening. His morning ration of grain was mixed with a can of condensed milk. A friend of mine who was an expert in equine hoof care trimmed the old black stud's long hooves. I spent a lot of time getting acquainted with Chicaro.

Within a month the grain and hay had made a big difference in Chicaro's appearance. He began breeding the mares. I kept him in the large paddock, and when a mare came into heat, I would lead her up to the fence to "tease" her to make sure she was ready to accept the stud. Next I turned the mare into the small two-acre lot to wait until I could lead Chicaro out for courtship. Since he had been pasture bred for so many years at the Salero Ranch when Dink Parker owned him, I thought it best to let him be free to court mares in the most natural way possible on the San Luis Ranch.

During the hours I spent with him, he gained confidence in me and would do simple things on voice command. I could tell him to go into his stall, and he would go. I would tell him to come out of his stall to me, and he would amble out to where I stood. I was amazed at the trust the old horse had in me. When he had courted a mare in the two-acre lot, I would hold out his halter and call him over, and he would obey. There was no chase, no hassle. I just put the halter on his head and led him back to his paddock. But there was one time that was different from the others.

Mrs. Thomison II, the little chestnut mare, came in heat. As usual, I led her to the paddock fence. Chicaro charged over, as was his custom, nickered, and snorted, and I saw that the mare was definitely ready to be bred. I turned her out in the two-acre lot and went back for Chicaro. I watched as he galloped over to her on the far side of the breeding pasture, courted her beautifully, and bred her. I held out the halter and called to him when the lovemaking seemed completed. Chicaro stopped and looked over at me for a moment. He lowered his head, shaking it as if he was saying "no" to me. I called again, but he didn't respond. He began courting the mare again, and after a while he bred her a second time. I called again and held the halter out for his head. The black stud walked over and put his head in his halter. I couldn't help feeling that with Mrs. T it was love; with the rest, only moments of lust.

By this time Chicaro was a magnificent stallion again. His black coat shone with brilliance, his ribs no longer showed, and his eyes had a new brightness to them. He also had a real friend in me, and I had a real friend in him.

During the early 1960s, many cattle deals and horse trades were finalized on the bar stools or in the booths in the El Dorado Bar in Nogales, Arizona. Chema, the bartender, witnessed a good many transactions from his position behind the bar. He didn't say much, but his friendly smile made customers feel welcome.

Since my principal business was a combination of cattle and horse trading, I came to know Chema, and one day I explained that I didn't believe in mixing booze and business. The arrangement Chema and I made insured that my trading instincts came from a clear mind. Should I order a Scotch and water while sitting with anyone, he would bring me just that, a Scotch and water. If I ordered a Scotch and soda, he would bring me a glass of ginger ale.

One late afternoon I had finished the chores and had gone to the house when the telephone rang. The man calling was one of Chicaro's owners from California.

"Come on down to the El Dorado and I'll buy you a drink," the man said over the phone.

"All right," I replied, "I'll see you in a bit."

As I drove along the highway, I wondered about the man's invitation. I had known him for several years, and this was the first invitation he had ever extended to me. It must be something to do with Chicaro, I thought.

I was happy to see Chema behind the bar as I entered, and I sat down with Chicaro's owners and another horseman from Bakersfield, California. Chema came out from behind the bar to the booth to take our order. The three others ordered Scotch and water. I ordered Scotch and soda. Chema moved his mouth in a half smile to let me know he knew what was going on.

The conversation rambled through three rounds of drinks, and I began to wonder why I had been invited to join these men. I could see that their drinks were starting to affect their speech. Then it dawned on me why they were taking so long to broach the subject. They were waiting for me to feel the effects of the Scotch. When Chema brought the fourth round, the man who had invited me for a drink looked across the table in my direction.

"Do ya wanta buy a horse?" he asked.

"Hell, I don't know," I replied. "I don't have any money to buy a horse."

A few moments of silence prevailed at the table. "What horse are you wantin' to sell?" I asked.

"Chicaro," he answered.

"I probably can't afford to buy him," I said.

Another a few moments of silence I asked, "How much do you want for him?"

"Twenty-five hundred," he answered.

"That sounds like a helluva lot for an old bastard that's old enough to vote," I replied.

The conversation drifted. Chema brought another round, and I could feel myself getting bloated from so much ginger ale. My companions were either feeling their drinks or had decided not to pursue the matter. I wanted to buy the old horse, but I was determined to get him on my terms. The other factor that drove me to attempt the consummation of the trade was their tactic of supplying me with enough booze that I would buy the horse on their terms. I also knew that if I had to pay $2,500 to keep my old black stud, I would. When I thought the right time had arrived, I tilted back my hat and laid out my offer.

"I'll give you a thousand dollars for Chicaro on a one year's note with no interest, and you transfer his papers into my name tonight."

"You just bought yourself a horse," he said.

I left the table and asked Chema for a piece of paper and a pen. I returned to my place and wrote out the note with its provisions and signed it.

"I assume you have the registration papers with you," I said.

"Here they are with the transfer application," he said, taking the documents out of his briefcase.

The horseman from Bakersfield interrupted. "Tell you what I'll do," he said to me. "Transfer the horse to me. I'll pay off your note in a year and write you a check for a thousand dollars right now. I'll pick up the horse next week."

"Sorry," I said, "Chicaro's not for sale."

I glanced toward the bar. Chema had been leaning toward the conversation. When he caught my glance, I saw a broad grin on his face. I ordered a Scotch and water on the next round of drinks. In a certain way I thought that I might have met and associated with so many coyotes that some of their traits were rubbing off on me.

Later Chicaro's former owners brought four mares from California to breed to the old horse at $200 per head stud fee. I told Chema about the mares coming to breed. "That horse is quite a gigolo," he said and grinned.

Two years later, Chema had moved his place of employment to another bar. I was getting ready to move away and went to town for an evening. The bar lacked clientele that night, so Chema and I chatted away. During the course of our conversation he asked me to lend him ten dollars to buy medicine for his wife. Without hesitation I pushed a ten-dollar bill across the bar, and we continued to converse.

I moved away shortly after that and over the years I completely forgot about the ten-dollar loan. Some ten years later I took a group of students on a field trip to the border. In a conversation with a man I had known during my horse-trading days, I learned that Chema had his own bar, El Tecolote, on the Mexican side of the border. During the course of the evening I decided to visit my old friend.

I walked into the Tecolote, and there was Chema standing not on the serving side of the bar but against the far wall, where he could watch what went on. He put on his famous broad grin when he saw me come through the door. We greeted each other with *abrazos,* and I filled him in on the changes in my life from horse trader to university professor. It was a joy to see Chema operating his own establishment.

In the midst of our conversation, he pulled a large roll of bills containing pesos and dollars from his righthand pants pocket. He peeled off a ten-dollar bill and handed it to me. "Here's the ten dollars I owe you," he said.

I held the bill in my hand. I could not remember. Finally it came to me. "Oh yeah," I said, "your wife needed medicine. You sure have a good memory, Chema. I had forgotten all about that." I put the ten-dollar bill in my wallet. Chema Estrella was no coyote.

Three years later my life had changed again, and I returned to Nogales after teaching at the university in Hermosillo for a year. I was almost broke. Things turned around in a couple of months, and one evening I decided to see if any of my old friends might still be around. I went to the bar where Chema had worked when I loaned him the ten dollars. There he stood behind the bar. Again the grin appeared, and we fell into conversation. I learned that the times had not been good for him financially, and he had lost the Tecolote. I stayed for a couple of Scotch and waters, talking with my old friend. When I got up to leave, I pulled out a ten-dollar bill, and left it on the bar.

"What's this?" he asked.

"Call it a tip," I said.

There it was again—the grin.

Among the many characters who frequented the El Dorado Bar was a man who talked a great deal but never seemed to have any livestock for sale or any desire to buy. He wore a cowboy hat, blue jeans, and a pair of high-heeled boots.

Jack O'Hara always wanted to be a cowboy. He was born and raised in cow country and had every chance to fulfill his dream. But there are some men, no matter their background or ambitions, who never become cowboys. Just because a man can saddle and ride a horse doesn't mean he is automatically a cowboy. O'Hara was one of these men.

During his life in Mexico and southern Arizona, Jack had a great number of jobs, many of which were in some way related to the cattle business. He could make most any person not acquainted with the cattle industry believe he was as real a cowboy as anyone who ever donned a broad-brimmed hat and a pair of high-heeled boots, because Jack O'Hara had refined his talent for telling tall tales into an art. So it was when he did a stint at selling real estate as a "ranch specialist."

There was a small ranch that Jack had listed for sale. By small I mean really small, with a patch of deeded land and an eighteen-head grazing permit in the Coronado National Forest. A large ranch might have a permit for a thousand head. In addition to being the selling agent, Jack became the "manager" of the place. Actually, he may have volunteered for the position so that he could have a roof over his head and the chance to tell everyone he was the manager of a cattle ranch.

Over the years he had acquired a large, red, bulbous nose from the many hours he had spent at bars and cantinas, telling the yarns of his exploits as a cowboy. Jack O'Hara could have been another Louis L'Amour if he had chosen to be a writer. He also had a substantial pansa, under which his pants hung in a sort of precarious limbo.

Carlos Dupré went to see Jack one day—out of curiosity more than anything else. When he arrived at the ranch house, Jack was nowhere in sight, so Carlos decided to go into the kitchen to see if there was any coffee to wait with and to sit down. The coffee pot was cold, so he heated the coffee on the stove, found a mug, filled it, and sat by the kitchen table.

Beside a scattered pile of mail on the table top Carlos spotted a medium-sized notebook and decided to look inside—again, out of curiosity. He had found Jack O'Hara's ranch log, with its daily entries. Most visitors might never have opened the notebook, but Carlos Dupré had to see what Jack had written.

He came to an entry: "Went to Thurber's Ranch to look at registered bull calves for a herd bull prospect." Carlos continued to read. Two days later the entry was: "Returned to Thurber Ranch and bot one registered bull calf for three hundred dollars." The following day: "Took delivery on Thurber bull calf. Turned bull calf out with the herd." Carlos read on until an entry three weeks later in the diary caught his eye: "Started roundup." For the following day the entry was: "Castrated Thurber bull calf by mistake." Jack O'Hara had changed a three-hundred-dollar breeding bull prospect into an eighty-dollar steer. Yeah, Jack was a real cowboy.

For one brief point in time, Jack O'Hara was my neighbor. After leasing a pasture that had once been under irrigation, he bought thirty crossbred steers and turned them out. The feed consisted of weeds and some dry grasses, but Jack thought it was enough to put a few pounds on the steers while he waited for beef prices to rise. The only water for the steers came from an irrigation well, and there was no trough for them to drink from. Jack made daily trips to the pasture, pushed the electric switch on the large pump motor, and filled a leaky ditch with water. By the following day the ditch was dry again.

This went on for a month. In the meantime, Jack bought another twenty head of steers even though his pasture didn't have enough feed for the original thirty steers. Soon after, he came to the house and informed me that the power company had cut off the electricity to the irrigation pump and he needed to move his cattle to water.

I climbed into his pickup truck and he drove into the pasture. His plan was to move the steers to a nearby corral, where he would load them onto trucks for transport to the livestock auction in Tucson. I told him that I thought it might be difficult to round them up with a pickup, but he insisted that he had done it before. He managed to drive over the rutted pasture, where irrigation borders had once directed the flow of water over the field. The steers were at the far end. When they saw Jack driving toward them, the steers began to move. I thought maybe Jack might be right after all. One steer, spookier than the rest, decided to leave his companions. Jack grabbed his rope, stopped the pickup, and opened the door. "You better stay inside or you'll lose them," I said.

Jack didn't listen to my warning. He stepped out of the truck, rope in hand, waving it at the recalcitrant critter. The entire bunch scattered in their panic.

"What the hell am I going to do now?" Jack asked. "I've got to get them to water."

"Either get a horse to work them or leave the gate to the pasture open," I said. "They'll smell the water down the road and go for it once they settle down."

Somehow I convinced Jack that my theory would work. He left the gate open, and that evening all the steers walked down the road and into one of my fields that had a full water trough. I sat counting them in the moonlight, and when they had all arrived I walked over and shut the gate.

A week later Jack and three of his friends came by to round up his steers. Instead of driving the cattle out the same gate where they had entered the field, Jack decided to push them out a gate on the far side. Six of the steers broke away from the riders, and from my vantage point it was quite a rodeo. One steer stuck his head between the second and third strands of barbed wire in the fence, trying to get back into the pasture. I walked over to chase him back out of the fence. Jack rode up, dismounted, and attempted to climb over the fence, but the crotch of his jeans became caught on one of the barbs in the wire. I was successful in getting the steer out and away from the fence, but Jack was hung up and helpless.

"Get me out here," he said.

"I'll hold down the wire, Jack, but I'll be damned if I'll unhook your britches," I replied.

It took a year before Jack found his six renegade steers, and the only way he recovered them was through the efforts of my other neighbor's cowboys.

Compañeros

*D*uring this time, Roberto Zúñiga, Carlos Dupré, and I were good friends. Roberto Zúñiga had attended the university when I did. Upon graduating, he operated his family's ranch south of Sásabe. The ranch provided him with income, status, and something to expand upon. Later, though, he moved his family to Nogales, Arizona, for schooling. Zúñiga had a mind full of ideas to make money, but like many "idea people," he had problems with the follow through. Not only did Bobby Zúñiga sell his own steers to gringo buyers, he also bought other steers for export. The first time I saw him after college was at El Club, across the line near Sásabe, while I was engaged in buying steers to put on the San Vicente Ranch.

Carlos Dupré owned a small ranch on the outskirts of Nogales that he had bought with the financial help of his father-in-law. He raised quarter horses and maintained a small herd of cattle. Carlos, Roberto, and I became friends, but their relationship was closer because their wives were good friends. I may have felt like an outsider, but I knew the reason why, and it made no difference in the escapades and adventures we pursued together. I had reached a point where the ranch operation was going smoothly, and I had a chance to do more than work, work, work. I felt the need to spend time away from the drudgery. The plans I had for the ranch could be accomplished when I got around to them. I wanted some camaraderie away from work. I found myself oriented toward Nogales and away from Tucson, where I had spent quite a number of my years and knew a lot of people.

Bobby Zúñiga wanted to raise quarter horses mainly for the prestige of it and the chance to meet the people of means who raised the horses. His problem was the border. It was difficult, if not impossible, to keep crossing horses back and forth across the line. His solution was to board his mare at the San Luis Ranch. That would be additional income for me, so I agreed. He didn't want to breed his mare to Chicaro; he wanted a Three Bars stud to impregnate his little darling. At the time, Three Bars had an excellent reputation for siring racehorses, a trait his progeny often passed on as well.

One Sunday morning, Bobby telephoned to ask if he could borrow my pickup and horse trailer. I told him he could borrow the trailer but I needed the pickup. An hour later he was at the ranch hooking his pickup to my trailer, getting ready to haul his mare to Mesa, just east of Phoenix, to a well-known stud.

"Why don't you come along with me?" he asked.

· "I've got too much to do around here, Bobby."

"I hate to make that damn drive all alone with nobody to talk to," he said. "I'll tell you what I'll do. If you come along for the ride, I'll buy lunch at the Cattleman's Club."

I had never been to the Cattleman's Club because I knew their prices were way beyond what I wanted to spend for any meal. But Bobby's offer was too good to pass up, so we loaded the mare into the trailer and started for Mesa on the Patagonia road. Before reaching Nogales, Bobby suggested that we stop by to see if Carlos would come with us. "Sounds good to me," I said.

Carlos came with us, though not before having a heated argument with his wife. Bobby promised to treat him at the Cattleman's Club as well. The trip to Mesa was fun because the three of us enjoyed bantering. Bobby spent an hour talking to the stud owner after unloading his mare. It was exactly noon when we arrived in the empty parking lot of the fancy restaurant. A sign on the front door announced that the Sunday opening hour was five in the afternoon.

"Well, I guess we might as well pick up some lunch somewhere, and go back to Nogales," Bobby said.

"Oh no," Carlos said emphatically. "You promised us the Cattleman's Club, and the Cattleman's Club it's gonna be!"

"We'll have to wait around until five," Bobby whined.

"We'll find some joint and wait until the Cattleman's Club is open," Carlos returned.

Bobby drove back onto the boulevard and found a bar within a half mile. We stayed there until five. Bobby and Carlos seemed oblivious to the fact that we were almost two hundred miles from home, a long drive with a belly full of booze. I decided that I would be the driver, so I limited my intake considerably.

As we entered the Cattleman's Club, I could see we were in an exclusive place—red velvet curtains, red leather seats, and huge menus, not only in size but also with offerings. I was startled by the prices. Bobby ordered a round of Chivas Regal. I was looking over the dinner selections when Carlos interrupted my thoughts.

"Why don't we have a chateaubriand for three?" he suggested flamboyantly.

"That sounds great to me," Bobby replied.

I motioned my approval, and closed the menu without noticing the price of their expensive-sounding selection.

The large oval platter came with the filet ringed with mashed potatoes—a truly sumptuous feast. I thought about my frugal New England father and how he would never consider such a meal. I wouldn't either. I had eaten the delicious cut from beef I had slaughtered and butchered myself, but to order such expensive fare in a restaurant never occurred to me. Actually, I had never been to a restaurant where chateaubriand was on the menu.

Always attempting to impress his companions, Bobby ordered a round of after-dinner drinks. The waiter brought the bill, on which Bobby scrawled his signature after writing in "15% tip." It didn't appear to me that he looked at the total. That act would be less than flamboyant.

I took command of the driving without any objections from Bobby or Carlos. They climbed into the back of the truck for a nap. I was approaching Florence Junction when I heard a banging on the rear window. I glanced around to see Bobby's concerned look and his hand up, signaling me to stop. His hat had blown away. After I maneuvered the pickup truck and trailer onto the shoulder he ordered me to back up. "Go walk for it," I said. "I am not about to back this rig on this highway."

It took him ten minutes to find his hat and return. I drove the remainder of the trip alone in the cab again. Bobby and Carlos slept in the bed of the pickup. I enjoyed the solitude.

A month later I was driving with Carlos along Grand Avenue in Nogales just before noon. Bobby spotted us and jumped out between two parked automobiles. I braked to a stop and he came to the passenger's side. "I got the bill from the Cattleman's Club today," he exclaimed excitedly. "It was eighty-five dollars! You guys are buying me lunch at the El Dorado!"

"Okay, Bobby," Carlos said. "We'll meet you there."

A lot of livestock people frequented the El Dorado Hotel Bar or the Montezuma Bar, on Morely Avenue, during the late fifties and early sixties. These were locations where we could socialize among people who were engaged in similar ways of life and businesses. Many cattle and horse trades were begun or consummated in those places.

Bobby ordered his lunch selection first, a steak. Carlos and I ordered the same. The conversation touched on an incident that had happened the week before. Bobby and Carlos had been in Bobby's office in Nogales, Sonora. Bobby was a ham radio operator, and he kept all his equipment there. He had contacted another ham in Detroit and requested a phone patch for a collect telephone call to me at the ranch. After picking up the receiver I heard the operator say, "I have a collect call for John Duncklee from Roberto Zúñiga. Will you accept the charges?"

I wondered what Bobby was doing in Detroit and why he was making a collect telephone call. Perhaps he was in some kind of trouble. "Yes," I replied.

The next sound I heard was laughter. Both Bobby and Carlos were in the background, laughing.

"What is going on with you guys?" I asked. "What are you doing in Detroit?"

More laughter. Then, "We're across the line at my office," Bobby said and laughed some more.

I suddenly understood what had happened. My two friends had pulled a good one on me, so I had to laugh with them.

We enjoyed the lunch, but the collect phone call joke kept going through my mind. I had to find a way to return the jest. When I finished my lunch, I excused myself. "I'll be right back," I said, winking at Carlos. "I have to see a man about a dog."

I left the table, walked out of the dining room toward the men's room, and kept going out the back door to my pickup. A few moments later Carlos arrived and sat down in the passenger seat, and we left. "I wonder how long Bobby will sit there before he realizes that we stuck him with the lunch tab," I said. Carlos and I were laughing so much that we were both at the point of tears. The next time we encountered Roberto Zúñiga, all he could do was smile and say, "You bastards."

Bobby learned how to fly and built an airstrip at the ranch, not an uncommon occurrence on Sonoran ranches. One time he had some two-year-old heifers at Felix Gómez, a ranch near Hermosillo. He wanted me to fly down there one day to help select the best of the animals for the ranch herd. The cut backs would go to the feedlots in Mexicali. Female cattle were not allowed to cross into the United States at that time, but beef carcasses could be exported. Bobby came up with the idea of fattening those heifers that were not good enough for replacements in his cow herd, slaughtering them, and shipping the carcasses to some U.S. meat company.

We took off from Nogales International and flew south without checking with Mexican authorities because our destination was a dirt strip far to the south. We had a low cloud ceiling, and Bobby climbed above the clouds. I always enjoyed flying above the clouds. It gave me a feeling of freedom, above the earth's problems, away from everything pending.

"I think we are lost," Bobby said after a half hour of silence except for the drone of the engine.

For some reason his statement didn't bother me. I calmly directed my eyes out the side window to look for a hole in the clouds. Five minutes went by, then ten.

"Three o'clock, Bobby," I said. "Bank her over and down. It looks clear below, but it's a damn small hole."

Bobby flew the aircraft into a downward bank perfectly through the opening in the clouds that I had spotted. There was a moment of hope and concern as we passed through the low ceiling. There might have been a mountain top waiting for us, but when we leveled out we were right at the end of the Felix Gómez strip! "Goddamn," Bobby said, and we landed.

Selecting replacement heifers is an important task in a cow-calf operation. The heifers chosen to join the cow herd should be better quality than the older cows to insure an improvement in the herd. Zúñiga looked at his cattle with an eye for the U.S. market and bred them to the best Hereford bulls he could afford. The former stocky, low-to-the-ground confirmation for beef cattle had changed to a demand for longer-loined cattle, so Bobby changed his breeding program to be able to sell at top prices. He also wanted longer-legged cows that could cover more country away from water holes and utilize his rangeland more efficiently.

We went into the largest corral, where the heifers waited. Three vaqueros on horses also waited to begin cutting individual heifers out of the herd. Another man tended the gates to two smaller corrals. If we decided a heifer was good for a replacement, he opened one gate. If we decided a heifer should be culled and sent to the Mexicali feedlot, he opened the other. Since I was well acquainted with the herd, Bobby asked my opinion on the heifers. We disagreed only twice. The heifers were his, so he made the decision.

By afternoon the clouds had disappeared. When we had finished separating the heifers into two corrals and were ready for the return flight to Nogales, there wasn't a cloud in the sky.

Several months later, Bobby told Carlos and me that he was going to have a local veterinarian check the replacement heifers for pregnancies. His plan was to fly us to the ranch on a Friday afternoon, spend the night, work the cattle the following day, and fly to Pitiquito on Sunday to watch a matched horse race, in which two horses run against each other. I couldn't resist such a weekend. Saturday would be a workday for us, but Sunday sounded like pure fun.

Late Friday afternoon we met at the airport and flew to Bobby's ranch. The veterinarian couldn't seem to stop talking during the flight. He may have been nervous about flying, but we discovered that evening that talking was his favorite activity anyway.

We enjoyed an excellent breakfast of *huevos rancheros* and beans early Saturday morning, and the vaqueros had the heifers ready for their examinations when we reached the corrals. To check for pregnancy, the vet had to insert his hand and arm through the anus of the heifer and feel the ovaries through the intestinal wall. This was done while the animal was restrained in a squeeze chute. The vet positioned himself at the rear of the chute while Bobby and one of the vaqueros worked the chute. Other men on horseback moved the heifer through the approach chute to the squeeze. I was mounted and in the middle of the corral. If a heifer was declared pregnant, I would cut her into the corral on my right. If she was not pregnant, or "open," I drove her into the corral to my left.

The operation went smoothly. As the heifers moved into the squeeze chute and were restrained, the vet shoved his arm in and felt around for a moment. "Thirty-one days, . . . forty-four days, . . . thirty-seven days, . . . open." It seemed a bit comical to hear the man call out such supposedly accurate estimates. Even the vaqueros who didn't speak English were amused to the point where they drove a two-year-old steer in with the heifers. When the steer jumped into the squeeze chute, the veterinarian shoved his arm up the only orifice the steer had in his hindquarters.

"Open," he said.

"If you'll lift up that heifer's tail, you'll see why she's not pregnant," Bobby said and joined the entire crew in raucous laughter.

When the steer was turned loose from the squeeze chute, he was angry and ran straight for my horse in the middle of the corral. I had been laughing with the rest of the crew and didn't realize that the steer was heading straight for me.

In seconds the steer had hurtled toward my horse and had gone between his fore and hind legs, striking my right boot with a horn in the process. The horse was as startled as I was and jumped. Somehow I stayed aboard, but my right foot hurt. I didn't say a word about it, but that evening I had trouble getting my boot off that foot. There was some swelling and a definite bruise.

The flight south to Pitiquito in the morning was smooth. As we approached the dirt strip outside of town next to the race-track, I noticed high-tension wires near the front of the bladed runway. Bobby was flying straight for them.

"Do you see the wires, Bobby?"

"Goddamn!" he exclaimed. "They weren't here the last time I flew in."

He dove the aircraft under the wires, and we touched ground just as a team and wagon started across the strip in front of us.

"Goddamn!" Bobby said again and pushed hard on the brakes. The left wing missed the tailgate of the old wagon by a foot. The old man driving the team hadn't seen us approach.

Bobby's foreman rode the winning horse. We had all placed bets on the winner and decided to drive into Caborca with some of Bobby's friends who had also picked the winner. I was just along for the ride because I was not well acquainted with the town. The car stopped at a *cantina* and we took seats around a large table in a patio. Tequila was ordered, and toasts to the winning horse and rider consumed the first round.

Shortly, another group entered the place and occupied another table at the opposite side of the patio. A *mariachi* band was playing inside the cantina. Bobby walked in and commissioned them to play for our table. Within minutes, another mariachi band had appeared, commissioned by the other group, who had bet on the losing horse. For an hour or so, Bobby tried to have his band drown out the other. The result was cacaphonic, but it was fun. I was happy to see that Bobby drank only one shot of tequila and ate a substantial lunch.

A month later Zúñiga telephoned to ask me to fly with him to Mexicali. He wanted to check on the heifers he had sent to the feedlot. I agreed to tag along. I wasn't particularly interested in looking at Bobby's heifers in the Mexicali pens, but I looked forward to flying with him.

Bobby followed the international boundary all the way to Calexico. As I looked out the window of the aircraft, I could see where the boundary fence was located by the distinct difference in the amount of grass cover on each side of the border fence. When I first started buying steers to put on Rancho San Vicente, I had observed that the ranges in Sonora had been grazed down more than those in Arizona. From the air, the difference was obvious. I didn't say anything about my observation to Bobby since his ranch was in Sonora and I didn't want him to take affront. A major factor causing overgrazing in Sonora near the border was the practice followed by many ranch operators of buying steers from below the tick zone for export. The steers were held on the northern ranges along with the cow herds. The longer they waited for crossing permits, the more they overgrazed the range. At times when the border was closed to cattle crossings, the problem was magnified.

When you're in the horse business and you come across a good load of hay at a reasonable price, you are inclined to buy that load whether there's room in the barn for it or not. I was lucky the morning a friend telephoned that he had encountered a man selling loads of Yuma hay for $125. A quick calculation revealed that the large, three-wire Yuma bales would cost me a little over a dollar a bale, a definite bargain.

I drove to my friend's place off the Tucson—Nogales highway to look at the hay and talk to the man selling it. I had no idea why the price was so low when I saw the top-quality hay on the semitrailer.

"Is that the price delivered and stacked in my barn?" I asked.

"I'll deliver it, and you can pay my two helpers ten bucks apiece to stack it," the short, stocky, red-faced man replied.

"Sounds good to me," I said. "I'll pay you and your helpers when it's in my barn."

I was very happy at the prospect of having a barn full of that excellent Yuma alfalfa at such a bargain. Four hours later I wrote the man a check and paid his helpers out of my pocket.

Later in the day I stopped by the El Dorado. I saw the hay dealer sitting at the bar with my friend who had telephoned me earlier that morning with the information about the bargain hay. I joined them and listened to the red-faced man tell about his renting a big house across the line. He continued with how he had bought a girl from the Zona Roja and installed her in his new residence. By his description of the house, we recognized it as a former bawdy house for wealthy Mexicans.

The hay dealer sold seven loads of Yuma hay during the week and then disappeared. A week later word flew around that a Yuma farmer had arrived in Nogales looking for the red-faced hay dealer and the hay. It turned out that the dealer had stolen the hay from the Yuma farmer. That explained the bargain loads. The sheriff couldn't find the hay thief, the farmer returned to Yuma, and I had a barn full of alfalfa hay.

Luis Castillo was one of my neighbors across the Santa Cruz River. He owned a small ranch and had a small herd of cattle. Some of the other neighbors in the vicinity had forewarned me that Luis couldn't be trusted and that I might find tools missing on occasion. But I wasn't inclined to take idle gossip too seriously. I wanted to discover for myself whether I could trust someone.

I was sitting on a cottonwood deadfall one morning, contemplating what I might do with the fields, when I saw Luis walking toward me from the river. We greeted each other and introduced ourselves in Spanish. I learned that Luis had very little command of English. He told me about the scant grazing conditions on his *rancho,* so I suggested that he could turn his cows into my field by the river. I had no immediate use for the area because I had no cattle, and the horses were in paddocks. He invited me to his house for coffee.

The following day I drove over to Luis's place and enjoyed his hospitality and conversation. I came back with the feeling that he would be a good neighbor and that those who had spoken against him probably had never made much effort to understand the old man.

For the next month I became fairly well acquainted with Luis Castillo. We shared coffee and conversation, and I enjoyed listening to his stories about the area and what times were like when he was younger. When I went out to the field with the cottonwood deadfall to sit and think, I felt good about seeing his cattle in my field. I felt good about being a good neighbor, and I felt good about Luis Castillo as my good neighbor.

It was late morning one hot, sunny day. I had finished building a fence along the road to enclose the east pasture and had walked to the cottonwood deadfall seat to think about how I would make enough income to live on the San Luis Ranch. Suddenly I heard two shots that sounded as if they came from a large-caliber weapon, maybe a thirty-thirty or a forty-five revolver. I didn't give the sound much thought and continued to sit on the deadfall, thinking. Perhaps a half hour after I heard the shots, I saw two sheriff's cars move swiftly along River Road and turn in at the entrance to Luis Castillo's place.

That afternoon I drove into town and heard what had happened. Luis had been shot to death by his neighbors to the north. I was shocked at the news. The neighbors, a father and son from Oklahoma, had built a house on the parcel of land north of Castillo Ranch. They claimed that their common fence line was mistakenly located several yards inside their boundary and were in the process of moving the posts to where they believed the legal boundary was. Luis had walked up to where the father and son were busy moving the fence and told them in Spanish to stop what they were doing. An argument ensued. Luis supposedly threatened to go home and return with his rifle. The Oklahomans went quickly to their house and returned with their weapons to the fence line. When Luis came walking up the hill with his rifle, they gunned him down with two shots. ·

The two killers were brought to the courthouse and went before the superior court judge in his chambers. After hearing their side of the story, the judge advised them to leave the state and never return. The judge claimed that he feared that Castillo's son would kill the father and son to revenge Luis's death. Two days later their house was sold, and they were on their way back to Oklahoma. I am not a lawyer, but I have always wondered why the two killers were not tried for killing my neighbor. Was Luis Castillo bushwhacked? Did he pose an immediate threat to the Oklahomans? I still wonder about the justice of turning those men loose to leave the state. I still wonder, if the tables had been turned and Luis had done the killing, whether the gringo judge, in his chambers, would have called it self-defense and sent Luis back to Mexico, or make him stand trial for murder.

I had occasional help with the ranch work. I didn't look for helpers, they just seemed to appear from time to time. The first to come walking up from the river was José. He was from Sinaloa. He walked in one midmorning and asked if I had any work for him. I thought about it for a moment and decided that it might get things done faster if I had someone to help. I showed him an outbuilding where he could sleep and found a mattress for a bed. I also found some cooking utensils that he could use and went to the kitchen for a food supply to get him started.

I told him that I would pay him the going day-rate for labor, which was eight dollars. The day-rate for Mexican nationals was three. I did not agree with that type of human exploitation. I also felt no remorse at hiring an illegal alien because a time or two I had gone to the park in Nogales, Arizona, to try to find someone to work. The men sitting in the park refused my offers, saying that they were making six dollars a day from the government by being unemployed.

Another exploitive practice by some was hiring illegal aliens, reaping the benefits of their labor and then turning them in to the Border Patrol without paying them. This practice goes beyond the realm of coyote and enters the criminal arena.

José had never driven a tractor, so I taught him how to operate the old Farmall I had bought with the ranch. I wanted to disk two of the fields to plant Sudan grass for summer pasture. The field nearest the road and the well had a slope to it that needed terracing, but I had plans to put in a sprinkler system eventually to remedy the situation. Because I didn't want to spend money on terracing or installing a sprinkler system at the time, I decided to attempt to irrigate the field in the same manner that I had seen hillsides irrigated in Wyoming.

I drove the tractor over the field next to the road so that any passing Border Patrol officers wouldn't see José. When I finished the disking, I began irrigating for a moist seedbed. I had made a ditch along the top of the field next to the well. I planned to fill the ditch from the well and make outlets as needed to spread the water over the field. My plan proved that irrigating hillsides in Wyoming was far different from irrigating a slope of sandy loam in Arizona.

José had been working with me for nearly a month when we began the attempt to irrigate the sloping field. He was down-slope, trying to spread the water over the field by making small diversion dikes with a shovel. Suddenly the water surged through one of the outlets, making a gaping hole in the ditch. I thought I could repair the damage without shutting down the pump, but the pressure was too much for one man with a number 2 shovel. I called to José for help. He had gotten to within twenty yards of the broken ditch when a Border Patrol car came slowly up the road and stopped at the well. Dammit, I thought, why did they have to show up right now?

I knew what was going to happen, so I went over to the pump and turned off the switch. The two "chiles verdes" (so named for their green uniforms) asked José the usual questions and took him into custody.

"He has some things at the house," I said, "and he has wages coming to him." The patrolmen cooperated with me and waited until I had retrieved some clothes and a valise I had given José. Then I handed José a twenty-dollar bill. I owed him sixteen, but I wanted to give him some extra. "You sure pay that guy well," one of the patrolmen said.

"Go out and grab that number 2 shovel and I'll pay you the same," I replied.

That afternoon I drove into Nogales, bought a carton of cigarettes, and gave it to José at the Santa Cruz County Jail. I wished him well and told him that if he managed to get across the line again, he would have a job anytime. I never saw José again.

Between running the horse business and other activities, I continued buying Mexican cattle. I wasn't using my own money, however. I bought the steers on order for a commission of two dollars a head plus expenses. The demand varied greatly, so order buying was just a periodic business for me. Under normal circumstances, the business of order-buying is seasonal because cattle are usually shipped in either spring or fall. The tradition is based on a number of things, but the most important is that the seasonal shipping dates come after the two growing seasons. The fall shipping comes after the summer rains have nurtured the range and the calves are in top condition and weight for sale. Should the summer rains not come, the weights are lower than normal and some cowmen hold their calves over until spring, hoping for a winter rainy season. This practice has sometimes caused overuse of the range unless the cowman reserved pasture for this purpose. The spring shipping can include both yearlings and calves, depending on the grazing conditions and the rancher's decisions. Across the line in Mexico, the same holds true near the border because ranchers in northern Sonora use many of the same management practices as those in southern Arizona. A rancher with only a few head to sell cannot necessarily afford to hold over cattle from one growing season to the next. Similarly, the cattle traders must sell as soon after buying as possible because many traders do not have enough rangeland to carry their "trading cattle" for an extended period of time. The system may seem complicated, but when one knows that weather fluctuations and the resulting range conditions, it becomes simpler to understand.

I had two clients in Kansas who wanted Mexican steers. One owned an auction, the other wanted the steers to put in his feedlot during the winter. I made several trips, some in the vicinity of Hermosillo, to try to find the kind of steers these men were looking for. The auction owner wanted three hundred head, but he wanted them uniform. To fill that order I had to look for a cow-calf operator who raised Herefords, or "Improved Mexicans." The other man wanted light calves that would put on efficient gains in the feedlot. He wanted four hundred head and would be happy to have crossbreds as long as the price was right.

I made several trips around the northern part of Sonora, looking for the steers that would suit my two clients. Most of the Improved Mexicans had already been sold, but I located two bunches of crossbreds that looked like they would fill the bill for the feedlot operator. The asking prices were higher than the man wanted to pay, however, and I couldn't buy the cattle for less. I telephoned him from Hermosillo to inform him that he would have to pay more per head, but the steers I had looked at seemed heavier than most and would therefore be worth more money. The man needed educating about Mexican steers. I reckon I wasn't that good a teacher on the telephone, because he didn't want to buy them.

I went to the Globe area in Arizona to look at some top-quality Hereford calves for the auction man. I rode through the cattle for two days because the country is rough and I wanted to get a good look at what I was trying to buy. The cattle were in fairly good condition, but I knew that working them out of those mountains would shrink them down considerably. The rancher planned to wean the calves at shipping. To my way of thinking, the cattle were a good buy. I telephoned the auction man, explained about the weighing conditions, and gave him my estimate of how much they would weigh off the scales.

"Get 'em for three cents a pound less and I'll take 'em," the auction man said. I knew that the rancher was very firm on his price. I suddenly decided that I wasn't cut out to be an order buyer.

"If you want these calves, come down to Arizona and see if you can find a better deal than I've lined up," I told him. "Otherwise I suggest you find someone else to buy cattle for you."

That was the end of my efforts to become an order buyer. I just didn't have the contacts, and it would be years before I did. As I drove back to Nogales, I decided to concentrate on quarter horses and, if I could grow enough pasture, to buy small bunches of old cows to graze and put on weight.

I began to get concerned about producing more income to support the horse operation. I had done some figuring, trying to see what it would take to make a profit. It costs as much to raise a top-quality foal as it does to raise one of lesser value. I concluded that I would either have to invest in better-quality mares or find outside work.

An opportunity for outside work came my way. A man who had attended the university during the time I was finishing my degree announced that he was planning to run for county tax assessor. He said he needed help with his campaign, and if he won he would hire me as an appraiser. I didn't know beans about political matters, but the idea of a steady income attracted me enough to tell him I would try my best with what little knowledge I had. He won the election and I became the field appraiser.

The position meant discovering improved real estate that hadn't been put on the tax roles, and there was plenty of it in the county. I spent a year and a half finding a multitude of buildings that had been neglected by previous assessors for whatever reason. It was interesting work, and I came to know many county residents.

Tubac had begun to evolve into an art-and-craft community, and a man from New York had built a shop to silk screen his unique southwestern designs onto high-quality fabric. I drove to the village to measure the structure for appraisal purposes and was impressed by the man's venture as a definite contribution to the community. I measured the exterior walls, including the living quarters, noted the materials, and discovered that the back doors to the building had not yet been installed.

"Since you have not completed the building," I said to the man, "I won't put it on the tax rolls until next year." This was a common and legal procedure. The man thanked me for the consideration and offered me some fabric for my wife. "No thanks," I said. "This is Arizona. You don't have to give anything to the tax appraiser here like you might have to in New York."

I happened on another source of income at the local radio station in Nogales. For about a year I was the "Country Reporter." Every morning I arrived at the station at 6:30 to go on the air with "The Country Report." I sold all the advertising, keeping 50 percent of the charges for myself as my pay. Before driving to the small building that housed the office and broadcasting equipment, I always stopped at a restaurant to buy a copy of Tucson's morning newspaper, the *Arizona Daily Star*. Part of the program was to give the livestock prices, which I read from the market section of the newspaper. The station didn't have anything as sophisticated as a ticker tape, so the *Star* was my only available source. There was also music from the turntable, which was operated by the engineer because I didn't have an FCC license. Of course, I didn't know how to operate the equipment anyway.

One particular morning will always remain etched in my

memory. I awoke with a definite intestinal disorder. I wondered if I would make it to the radio station without stopping along the highway. I didn't, but I did finally make it to the studio.

"Give me the mike with the long cord," I said to the engineer hurriedly. He handed the microphone to me and plugged it in as he opened the transmitter for the day. I charged into the bathroom and sat down. "Good morning," I said as usual. "This is John Duncklee with your country report. And now for a song to welcome your day." I generally started the program with the cattle market.

The engineer started the music, and my intestinal tract exploded. When the song finished, I said something about being glad that I was on radio instead of television, read the cattle market, and announced that there would be more music. Somehow I managed to talk about the sponsors during the half-hour program. When it was over, I drove back to the ranch and found the Kaopectate bottle.

I had acquired another job at about the same time: advertising salesman for a quarter horse magazine located in Amarillo, Texas. *QHB Magazine* had been started by a handful of quarter horse breeders who were disenchanted with the American Quarter Horse Association, or at least with its monthly publication, *The Quarter Horse Journal*.

When I was showing two mares at the annual show in Prescott, I met the editor of the rebel magazine. We talked for a while, and he hired me as western states representative to sell advertising. I was surprised that he issued me credit cards for air travel and phone calls. In addition to giving me a commission on sales, he agreed to run a monthly half-page advertisement for Parker's Chicaro, who was at stud for two hundred dollars at the San Luis Ranch, guaranteed live foal.

The men who were trying to compete with the *Quarter Horse Journal* must have had bundles of money. The editor purchased a new Thunderbird sports car, and all the advertising representatives traveled to Amarillo for four days to telephone everyone who owned quarter horses to sell them ads or subscriptions to the magazine. We sold some advertising, but nobody knew whether the campaign had been successful. The magazine went out of business about six months after that, and I received a letter from some Texas lawyer informing me that I owed several hundred dollars for the half-page ad for Chicaro. I wrote to the lawyer, told him about the agreement I had with the editor, and enclosed the credit cards. I had sold several pages of advertising but had never received a commission check.

During the time I was associated with the magazine I had an experience that is impossible to forget. During the telephone ad campaign the editor asked me to sit down in his office to listen to an idea. He had been thinking that a yearling quarter horse sale similar to the Keenland Sale for Thoroughbred yearlings would be successful. I agreed that he had a good idea, and the more I thought about it, the better it seemed.

Rillito Racetrack in Tucson had once been the quarter-horse racing capitol, but Thoroughbred racing had become more popular. As a result, the track at Los Alamitos, California, had become to quarter horse racing what Rillito had once been.

I was interested in anything that might arouse interest in quarter horses in Tucson again, so I suggested that the sale be held at the track if I could arrange it with the current owner of the track. The *QHB* editor offered to pay me $2,500 if I could nail down an agreement with the owner and would manage the sale. I knew the task would require many weeks of work if the project was to be successful, but the idea of reviving interest in quarter horses in Tucson made the proposition seem worthwhile.

The sale entries had to be by AAA-grade running sires, or out of AAA running dams. There was a fifty-dollar nominating fee to cover the cost of inspecting every nomination. The sale would be limited to 150 yearlings. It promised to be a first-class sale with first-class horse flesh.

I telephoned the owner of the track, and drove to Tucson for our appointment. When I explained what we intended to do, he became excited. He could see that a sale of that quality would enhance the racetrack. He offered all the track facilities for nothing and added that he would buy advertising in addition to what we had planned. I couldn't have asked for a better deal than that. I drove back to Nogales filled with elation and called the editor in Amarillo with the good news.

The following month's issue carried the first announcement of the sale and called for entry nominations. I kept in regular contact with the office, and within two weeks after the magazine went out to the subscribers, fifty yearlings had been nominated. A week later I received a strange telephone call from a man who sold horse insurance. The agent mingled with wealthy horse owners in several states. I had seen him at various quarter horse sales and shows. The telephone conversation began in a friendly tone.

"Who is handling the insurance for your sale, John?" he asked after he told me that he thought the sale was a great idea.

"I have no idea," I replied. "Insurance is not my department."

"I think you need to make it your department," he said. "Those guys in Amarillo don't really know what is going on."

"Like I said, it's not my department. I have enough to do without getting tangled up with insurance."

"Let me put it this way, John," he said. "If I don't handle the insurance, you probably won't have a sale."

"That's interesting," I replied to his threat. "Why don't you call the guys in Amarillo and tell them what you just told me?"

I hung up the receiver. I didn't know what to do except telephone the editor and report the threat by the insurance agent. I not only didn't want to get involved with the insurance, I also had no desire to have any contact with a man who threatened me on the phone. The editor told me not to concern myself with the matter, so I put it out of my mind.

The following week I was in Tucson on business, and as I walked toward the entrance to the hotel that livestock people frequented I saw the hotel owner and a politician from northern

Arizona leaving. The politician managed quarter horse sales in the northern part of the state, and it was common knowledge that he wanted to become "Mr. Quarter Horse Sale of Arizona." Seeing the two men together didn't mean anything to me at the time, but three days later the pieces of a sinister puzzle fell into place.

The owner of the racetrack telephoned to tell me that he was backing out of our agreement. I asked him to explain, but he said he couldn't talk about it over the telephone and for me to come to his office. I spent a sleepless night wondering what was going on.

The owner's reason for withdrawing his offer of the racetrack for the yearling sale both astounded and angered me. The hotel owner had threatened to boycott the racetrack with his many friends if our sale was held. I have never been able to understand how this conniving little weasel of a man could have any friends to influence when he was one of the most coyote-like individuals I had ever met.

The next piece of the puzzle fit perfectly. The hotel owner and Mr. Quarter Horse Sale of Arizona announced plans for a yearling sale to be held in the lobby of the hotel!

In our discussions about our yearling sale, we estimated an average price in the vicinity of $2,000 per head because of the quality standards we had insisted on. The yearling sale in the hotel lobby averaged $250 per head, and several of the yearlings were passed out of the auction ring by their consignors because of the lack of bidders. The sale turned out to be a gigantic flop and did nothing to increase interest in quarter horses in southern Arizona.

Situated as they are on the U.S.—Mexico border, Nogales, Arizona, and Nogales, Sonora, are heavily involved in international trade, both legitimate and otherwise. As I said earlier, the boundary is also the most culturally contrasting of any such boundary in the world. Much of what was common knowledge about smuggling in Nogales rarely went farther than the border until the area became a drug smuggling corridor from Mexico into the United States. Before that, there had always been an active smuggling trade going both ways, depending on the item being smuggled.

Smuggling appliances and electronic products into Mexico by American merchants was a matter of mordida, bribes to Mexican customs officials—no big deal. Clothes were also less expensive on the U.S. side than across the line. Untold numbers of Mexican women would buy several dresses in the U.S. department stores on the border, put them on in layers, and walk across the border to sell their contraband.

During the early sixties, Scotch whiskey sold for eighteen to twenty dollars a bottle in Mexico but could be purchased wholesale in the United States for a fourth of that. There was a lively business in smuggling Scotch whiskey by air into Mexico among a group of men who owned and flew aircraft with great skill. The most popular aircraft for a while was the Piper Comanche, though one of the smugglers, a Mexican citizen, flew a twin-engined Cessna. Everything except the oxygen equipment was stripped from inside the cockpit to make the most space available for cases of Scotch. Mostly it was Cutty Sark because that brand was the status symbol for middle-class Mexican Scotch drinkers.

The aircraft were loaded in the United States, mostly at one of the local airports, and then flown to remote airstrips throughout Sonora and possibly Sinaloa. After landing, often at night with bonfires to mark the landing strips, the cases were loaded onto trucks for transport to their urban destinations. The return flights often carried a variety of items made from silver. Thus the flying smugglers broke no U.S. laws, only Mexican. This business went on for several years. The Cessna-flying smuggler was caught by the Mexican customs officers, and the Main Man gringo decided to try a larger aircraft. He crashed and was unfortunately killed. After that the Scotch-smuggling business seemed to leave the area, and the people involved in it went on to other enterprises. They had certainly helped to make those times interesting.

The stock market was being good to me. I decided to sell some of my securities and begin paying off the $25,000 loan I had with the bank. Actually, instead of selling just enough to begin the payback, I sold enough to liquidate the loan entirely. However, I first paid just half so that I would have some cash if a good cattle or horse deal passed my way. A week after I went to the bank and paid off half the loan, I received a phone call from the bank manager.

"John, I have been thinking that we should sit down and make some sort of plan to pay off your loan," he said.

I was surprised. Obviously the banker had not looked at my file. "That sounds fine with me," I replied. "I'll be in as soon as I can get there."

I dropped what work I was doing and drove into Nogales with my checkbook. I sat down in the chair next to the banker after he gave me his banker's smile and handshake. Without another word I wrote a check for $12,500, and handed it to him.

"You don't have to pay that much right now," he said.

"Let me know how much interest I owe, and I'll write you another check."

The confused banker went to a file cabinet and pulled out a manila folder. He saw his mistake, but it was too late. I felt strongly that I wanted a banker who knew where I stood before suggesting a plan to pay off a loan that had been reduced by 50 percent the week before. I wrote the interest check and then transferred my accounts to the downtown Tucson office.

Paying the Bills

The horse breeding part of my business was proceeding as planned, but I knew that I had to derive more income from some source or I would deplete my capital. The job with the county assessor ended, the job as advertising salesman for *QHB Magazine* ended when the magazine folded, and I had tired of the radio program. I welcomed the offer of a man who wanted to lease the ranch's pastures. The only problem seemed to be that he had no cash and wanted to lease the place in return for terracing the farmland. I pondered the proposition for several days and decided to require that he put up some sort of performance bond to insure his compliance with the lease agreement. I had learned something about business by this time.

He offered to put up a dozen two-year-old heifers and allow them to graze the pastures he was intending to plant. I agreed. The heifers arrived, and we branded them with my brand, the O Bar J. I turned them out into a field I had planted to Sudan grass. The mares had been there for a while but not long enough to clean out the field. I anticipated the terracing and planting of the other fields, but the lessee went to work cutting mesquite trees for charcoal processing across the road. He said that he had to sell the wood in order to raise the money to terrace and plant the fields.

I soon realized that there was something wrong with the deal when the grass ran out in the Sudan grass pasture and the lessee had done nothing in the way of what he had agreed to. I moved the heifers to another pasture and began feeding them hay. After two months of watching the trees cut down, sawed into fireplace logs, and stacked into one-cord piles, I decided to confront the lessee. He maintained that he had a buyer for the wood and planned to begin terracing the fields as soon as he received money for it.

I learned that the man was spending a bundle of money in the bars and was living in the El Dorado Hotel. He was also involved in an affair with a married woman, and he had no buyer in mind for the mesquite he had cut in order to make more tillable land on the San Luis Ranch. I knew that it was time to make some sort of move either to force him to fulfill his agreement or take steps to have him evicted. The two-year-old heifers were beginning to cost a lot of money to feed, and I was paying the feed bill.

I started eviction proceedings, and within two weeks the man had signed the necessary papers that returned the San Luis Ranch to my complete control. I did not enjoy the proceedings and wished he would have tended more to business than to cavorting around and spending what little money he had finagled from his father. I guess I hadn't yet learned all I needed to learn.

The two-year-old heifers started calving, so I was busy keeping a close watch on them in case they had problems. I made sure that any newborn calf had its navel swabbed with Smear Number 62, a screwworm remedy. I also watched them closely for any problems in calving.

At about this time, an acquaintance from Saint David, without asking me, left a Rhodesian Ridgeback male dog chained to one of the mesquite trees in the yard. I can only assume that he thought the dog would be a good mate for my Ridgeback female, Lennie.

I didn't like seeing the dog chained to the tree. I approached him cautiously. He seemed friendly enough, so I reached down to stroke his head. I noticed that the chain had been around his neck for a considerable time. He continued to be friendly and seemed to welcome my attention. After ten minutes or so of making friends with each other, I removed the chain, and the dog followed me around the barnyard as I completed the evening chores. I kept him in the back room next to the kitchen for the night, and in the morning he again followed me around during the chores after enjoying his breakfast.

I walked out into the pasture to check the heifers and found that one had calved during the night. I approached the calf, grabbed it, and applied the screwworm medicine to its navel. As I walked back toward the barn, I heard the calf bawl. I looked in the direction where I had left it and saw the Ridgeback on top of the calf with his jaws around its throat. I ran toward the scene, grabbed a rock from the ground as I ran, and threw it as hard as I could at the dog. I guess my having been a baseball pitcher paid off that day because the rock hit the dog in the ribs, and he turned the calf loose. I went over to inspect the calf as the dog sulked away. The hide wasn't broken, but the calf was stunned from the experience. I rubbed its neck until it recovered from the shock and got to its feet to trot over and join its frantic mother.

I called the dog to me and walked to the house where I kept my rifle. There was nothing else to do but put the fellow down. I couldn't have a calf killer that might even become a foal killer on a ranch devoted to raising livestock.

The Southern Arizona Livestock Association show and sale was an annual event that most cattlemen attended. I enjoyed looking at the show-conditioned cattle, and passing the time with acquaintances I might not see otherwise.

A livestock veterinary supply salesman was always in attendance, along with his booth displaying the latest in vaccines and drugs that his company had developed. He was, of course, friendly and hospitable, being a born salesman. His hospitality went as far as several bottles of whiskey hidden behind the booth, and he was very generous with the libations poured into soft-drink cups to disguise the real contents. I chatted with the man for a while before going to the sale ring to watch the cattle auction.

I sat in the bleachers for fifteen minutes without seeing anything I wanted to buy, but then three Brangus heifers were brought into the ring. They were nice-looking cattle, and they all looked like they were carrying calves. The auctioneer read their registered bloodlines and started his chant. The livestock loan officer I had dealt with was acting as the clerk of the sale. I caught his eye and with sign language asked if he would approve these heifers for a bank draft. He made an OK sign with his fingers, so I proceeded to bid for the cattle. I was surprised to find myself the buyer for $1,100. Getting three registered Brangus heifers, possibly bred, for that amount was a definite bargain.

I went out to my pickup truck, took a bank draft out of my briefcase, and returned to pay for my purchase. The next step was to drive to Nogales for my horse trailer. I hoped that I could squeeze all three into a trailer built to hold two horses. As a precaution, I brought two boards and some baling wire to make sure they couldn't escape over the tailgate.

The sale was over by the time I returned, but a couple of men helped me get the heifers loaded, and I headed back to the ranch with them. I put them in the corral I had built across the road from the farmland, and the next morning I branded them with the O Bar J. Two months later, one of the Brangus had a bull calf. Three months later I sold the twelve Herefords and the three Brangus with their calves for a nice profit.

Cabezón

fter selling out the O Bar J herd, I moved to Tucson. Then I set out on a trip that was supposed to take me through Arizona and New Mexico. I wanted to find a new place to live and raise horses and cattle. The trip was also a way that I could travel by myself and be by myself to think by myself. This happened before I went to Mexico to buy steers, and before I became partners on the cotton farm.

My first stop was Nogales. I knew one man in the town, Paul Bond, the well-known western bootmaker, so I stopped in to see him. Paul directed me to Joe Escalada as someone who could tell me about the area better than anyone else. After talking with Joe for a while, I decided to spend the night in Nogales. After another day roaming around the border country, I decided that I need not continue the exploratory trip. I had found where I wanted to live. I had also begun a friendship with José Escalada that would endure until his death.

I made frequent trips to the new Escalada Ranch Supply store on the Tucson—Nogales highway after moving to the San Luis Ranch. There was a small round table in one corner next to a coffee-making machine that always held hot brew. It was always a pleasure to sit down and talk with a number of people, many of them old-timers, who also enjoyed the Escalada hospitality.

The Escalada brothers—Luis, Manuel, and Joe—continued the business their father had started upon arriving in Nogales. José had been born in the vicinity of Santander in northern Spain. Manuel was in charge of the grocery end of the business, Louis handled the liquor and tobacco, and Joe, also a cattleman, operated the ranch supply part. Louis was the politician of the group and had been appointed as a state highway commissioner. His stories seemed endless. Joe smoked or chewed rough-looking cigars that he had to send for out of state. Joe had an anthology of stories about his youth, a six-year visit to Spain, and cattle people.

Among the coffee drinkers I came to know was Les Wooddell, who had ranched on the Yaqui Indian Reserve near Guaymas, Sonora. He carried his nickname, Cabezón (Big Head), wherever he traveled in southern Arizona or Sonora. As a youngster he had come with his family to Arizona from Minneapolis, where he had been born in 1890. After attending schools in Phoenix and Tucson, Cabezón played football on an early University of Arizona team. In 1906 he witnessed the Colorado River overflow its banks to fill the Salton Sea. He worked for the Arizona Livestock Sanitary Board, registering all the Papago Indian brands. He managed the Arivaca Land and Cattle Company for Bogan and Bernard, and then managed Jarillas Cattle Company holdings for Jack McVey. In 1919 he went into partnership with Joe Wise on La Arizona, a Sonoran ranch southwest of Nogales that used the JW brand. Joe bought out Cabezón in 1931, and Cabezón went into partnership with his father-in-law, Gene Sykes, on the Pozo de Crisanto and Carbó ranches near Horcasitas, Sonora. The partnership was short-lived because of "bad neighbors," and they sold their ranches to the former governor of Sonora, Francisco Elías.

The early 1940s found Cabezón negotiating grazing leases with the Yaquis, a significant accomplishment. He was the first man to import Brahman cattle, and later Charolais, into the state of Sonora. His crossbreds became some of the finest in Sonora. A bronze sculpture of him is located at the feedlots in Carbó, near Hermosillo. Ranching next to the Yaquis had its adventures.

Early on, Cabezón was riding through his range and came upon several Yaquis butchering one of his animals. He dismounted, walked over to the scene, and boned out a tenderloin. Without a word he tied the meat on the back of his saddle, mounted, and rode home. "I told them in that manner that I knew they were stealing one of my steers," he told us one day at the coffee table in Escalada's store.

Once he discovered that a good number of his horses had been stolen. He saddled the wrangle horse and rode up the mountain to Yaqui headquarters. Without looking to the left or right, he rode straight to one of the village governors and told him about the theft. The Yaqui was sufficiently impressed with Cabezón's bravery and presence that he told him that his horses would be in his corrals by the time he returned home. When Cabezón reached the ranch, every horse that had been stolen stood in the corral.

To complicate matters, the Yaquis were at war with Mexico during Cabezón's early days of ranching there. A young Yaqui arrived one night with a Mexican bullet in his shoulder. Cabezón drove to Guaymas for an army surgeon to treat the boy, but the doctor refused to travel so close to the enemy and gave him instructions for removing the bullet and dressing the wound. Cabezón returned to the ranch, extracted the bullet, and kept the wound dressed until it healed. Only then did he send the young Yaqui back to the mountain.

He learned shortly that the boy was the son of an important Yaqui village governor. His act of kindness earned him eternal respect from the Indians. In 1960 the Mexican government informed Cabezón that he would have to vacate his leases with the Yaquis. The village governors came to Cabezón and offered to go back to war with Mexico so that he could remain there with his cattle even though the government was buying six thousand head from Cabezón to give to the Yaqui Tribe. Cabezón declined their offer but appreciated their concern. After the cattle were counted and paid for, Cabezón bought two ranches between Hermosillo and Guaymas, the Santa Margarita and the Mosco Bampo.

When the new crossing corrals were opened at Sásabe, Cabezón drove a group of us from Nogales to the party. When we were seated in his Studebaker, he turned slightly to make an announcement. "I'm the world's worst driver," he said, "and I'm very sensitive about it."

That gave us all a good laugh to start the trip with. As we passed through burroweed- and mesquite-invaded rangeland west of Arivaca he showed us where he had once hired off-duty Mexican soldiers to cut the native grasses for hay. That was during Pancho Villa's time. He stopped the car and looked out the window. "I wouldn't piss on it now," he said nostalgically.

The party thrown by Mike Knagge, whom many called "the mayor of Sasabe," was attended by just about every cowman and cowboy in southern Arizona. No invitations had been sent; there was a general invitation by word of mouth. Mike furnished the food, the booze, and the music for the party, which attracted people from both sides of the border. It was an occasion that those in attendance will never forget.

Les Wooddell, Cabezón, had the deep respect of anyone who knew him. He died in Nogales in 1966.

At the End of a Halter

Quarter horse shows were held all over the country in the 1950s and 1960s, and Arizona had its share. These affairs presented the opportunity to showcase the kind of horses a breeder raised. Some shows were accompanied by auctions, and the usual rule was that any consignments to the sale had to be entered in the show. Entering an animal in a show involved a lot of work before, during, and after the event.

Before appearing in a show, a horse had to be trained, groomed, and generally put into what was called show condition. Analyzing the judge could be important because there would be little benefit in showing a type of animal that a particular judge didn't like. The order in which a judge places the horses in most any class is only his opinion. Sometimes, however, his judgment might be swayed by other circumstances, as may have been the case at the Prescott show one fall.

The judge was a prominent quarter horse breeder from Oklahoma who had shown his own horses in Arizona for a number of years. He arrived in Prescott in his new white Cadillac convertible and attended the exhibitors' cocktail party and dinner, hosted by a major bank. After dinner there was dancing, and the judge tripped the light fantastic with a flashy-looking woman, well-known on the Arizona horse show circuit.

I went to my motel before the festivities had ended. The

following morning, as I left my room for breakfast and the fairgrounds, the judge and the woman walked from the room next door to the convertible. It was interesting to watch the woman receive two blue ribbons in two performance classes, reining and western pleasure, later that afternoon. I was glad that the two mares I had entered in the show were in halter classes.

These horse shows were tiresome, expensive, and boring ordeals for me. I may have sold a couple animals because I was present at one show or another, but I doubt it. I think I would have made the sales anyway. They seemed to be more like social events for the exhibitors than anything else, but occasionally they may have resulted in some good horse trades. The auctions that accompanied a few of the shows were another thing.

The American Quarter Horse Association's annual convention was held in Fort Worth, Texas. a year or so after I moved to Tucson. I was returning from Kansas with my wife at the time after hauling two mares to a stud. One mare belonged to her; the other belonged to a friend of mine in Tucson. I was hauling the two-horse trailer empty on the return trip. Fort Worth was not out of the way, so we decided to stop for the convention and its accompanying horse auction.

The convention was like any other convention, but the catalogue for the auction looked interesting. We went out to the fairgrounds to look over the consignments.

A bay mare looked like a good prospect, both from her confirmation and her bloodlines. After a discussion about the mare and how she would fit into our breeding program, we decided to bid to $2,500, if necessary, to buy the bay. I assumed I would be doing the bidding.

The following day we entered the large tent and sat in the front row of seats. The editor of the *Quarter Horse Journal* was the ring man on our side, and he worked in front of where we sat. It was common for men associated with various horse magazines to serve as ring men, taking bids from the audience. I was acquainted with the man and had introduced him to my then wife.

The auction began with a number of yearlings and two-year-olds going under the auctioneer's hammer. Finally the bay mare was led into the ring for everyone to see. The auctioneer started the bidding at $1,500. I waited. When the bidding rose to $1,800, I waved my hand to bid $1,850. I watched the ring man closely as the bidding continued. I judged from the his eye movements that whoever was bidding against me must have been sitting close by. Right after I bid $2,200, I looked to my left and saw my wife bid $2,250!

I rose from my seat and told the ring man to cancel all my bids and all my wife's bids. "That's a helluva way to try and sell a mare," I told him. I returned to where my wife sat and said we needed to leave the tent. We didn't buy the bay or any other mare that day, and I drove the empty horse trailer back to Tucson.

Three months later the two mares in Kansas were settled to the stud, pregnant, and I drove there and hauled them home. My friend's mare had done some racing in her youth, but the stud she was bred to was a cutting horse sire, not a father of racing stock. The breeding was an experiment.

One evening the following January, he phoned to say he thought his mare might be ready to foal. I went over to his place in the Tucson Mountains, west of the city. We walked out to the stall area and went into the mare's run to look at her. Her bag was full and her teats waxed.

"She looks ready to me," I said. "Have you seen her go down at all?"

"A couple of times, but I'm not sure if she's in labor."

We returned to his living room to give the mare a chance to do what she had to do. In an hour I suggested we go out for another look. The mare was down in labor with two yellowish hooves and a tongue showing. "There's your foal," I said. "Let's go back in and let her finish alone. Everything looks normal. We can check her out in a half hour."

When we returned, the foal had been born and the mare was in the process of expelling the afterbirth. It was a cold evening, and my friend worried about the wet foal being out in the weather. He ran to the house, returned with an armful of towels, and began drying and cleaning the little filly. "I think you should leave it alone," I said. "Get that little devil too clean and the mare might not accept it." He continued to rub the filly dry.

It didn't take long for the filly to start trying to stand, and after a few unsuccessful attempts she made it up to a wobbly, spread-legged standing position. The mare was at the far end of the run, taking no interest in her newborn. The filly swayed and wobbled over to her mother. The mare took a smell of the little one and moved away. "Dammit, she's not takin' the filly," my friend said with frustration and disappointment.

I walked into the run, grabbed the afterbirth from the floor of the pen and smeared it over the foal, rubbing it over its entire shaking body. I then went to the mare and led her to her offspring. The mare put her head down to the filly and began to lick the afterbirth away. As the filly searched for her mother's bag, the mare turned her body and nudged the filly toward her first meal. My friend and I went back to his house and drank a toast to the event.

My Last Horse Race

The men who had sold me Chicaro purchased a farm near Tumacacori. They had grandiose plans to become Mexican steer buyers in addition to race-horse breeders. They built a feedlot for cattle and corrals for their horses. I drove out to their place one day, curious to see what they were doing.

The parts to a feed mill were scattered around the feed-lot area, and a few good-looking horses nosed the hay in their mangers. I noticed a particular sorrel filly that I had seen shortly after she had been foaled. "She's quite an athlete," the older brother said. "We're training her for the Southwest Futurity."

The Southwest Futurity, at the Rillito track in Tucson, had become a popular race for two-year-old quarter horses. I was interested in seeing the filly run since I had known her almost since her foaling and knew both her sire and dam.

The Saturday of the futurity, I finished what chores I had to do as quickly as possible and drove to Escalada's. Joe cashed a fifty-dollar check for me, and I told him I was planning to use the entire amount to place a win bet on the sorrel filly.

"That's a lot of money to put on a horse's nose," he said.

"I know," I replied. "I generally don't bet more than two dollars, but I really like the looks of that sorrel."

I left Joe grinning through his cold cigar and drove to Tucson and the futurity. I found a parking place at the track and walked to the entrance. As I paid my dollar to go in, I heard the loud-speaker announce that the sorrel filly had won the Southwest Futurity and paid twenty to one! I had not only missed the race but also the chance to put my fifty dollars on her nose.

I found the brothers who owned her in the clubhouse. They were elated over the outcome of the race. I told them that I had arrived too late to place my bet. "Hell, John, I didn't have a dime on her myself," the older brother said. I thought it odd that he had less faith in his entry than I did.

I enjoyed raising horses, but I had no desire to get into the racing business. I was content to sell the offspring of my mares to people who wanted them. If they trained any of them for the track, that was fine with me.

The mare that I had hauled to Kansas for breeding foaled a filly. I was determined that she would be bred to a racehorse sire. I hauled her to Willcox to be bred by Rukin String, a famous running quarter horse I had seen in Tucson when he was two weeks old many years before. Rukin String had been bred and raised by Rukin Jelks, a well-known horseman who had much success in the horse business throughout his lifetime.

Rukin String's sire was Piggin String, a Thoroughbred. His dam was Queeny, a club-footed mare that had been a champion running quarter horse. Rukin String accomplished the same feat two years in a row. When I took Chicaro's daughter to him, Rukin String belonged to Harry Saxon in Willcox. The stud fee was $300, a helluva good gamble in my estimation.

A year later the mare foaled. The little stud colt looked good. As he grew, it was obvious that the colt had the makings of a real athlete. When he was six months old my friend in Rockford, Illinois, telephoned to ask if I had any horses for sale. I described the stud colt to him and priced him at $1,500. "I'll be out to get him in a couple weeks," the man said. It made me feel good that he trusted my description enough to travel all the way west to Arizona from Illinois. The colt was broken to lead, so there was no difficulty in getting him loaded. I weaned him at the moment he stepped into the man's horse-trailer.

Three years later I received a telephone call from the man in Rockford. He asked whether I had any more colts. I told him that I was out of the horse business but that I was curious as to what had happened to the Rukin String colt.

"Well," the man said. "I trained him and put him on the track. He's come in second twice."

"Hell, that's not bad," I said.

"But he's run first ten times."

"No wonder you're looking for another one of my colts," I said.

Carne Asada on the Hoof

*L*ife on the U.S.—Mexico border during the early 1960s was challenging, interesting, and pleasurable. The maquiladoras (the twin industrial plants), had not yet been instituted, so the population in Nogales, Sonora, had not yet exploded. Traffic across the line was enough to be bothersome at times, but there were not the entanglements on the streets seen in recent years. There was no drug traffic, or at least I wasn't aware of any. It was a unique opportunity to experience a cross-cultural situation, both social and economic. I not only had acquaintances on both sides of the border, I also spent as much social time in Mexico as I did in Arizona.

Once I was invited to the baptism celebration for the baby daughter of a good friend of mine. Mexicans celebrate such events with more serious parties than we do in the United States, so I looked forward to attending. I didn't go to the church for the baptism part but drove across the line to the home of the prominent businessman who was the godfather of the infant. With the honor of being the godfather went the honor of throwing a party for the adult guests. I arrived at the appointed hour of five in the afternoon.

The host's house included a swimming pool (empty) and spacious rooms. At the rear of the house a smaller structure served as a laundry room. The living room and dining area combined to accommodate a large crowd of well-wishers. I knew some of them and was introduced to others.

The beer was my favorite brand, Pacifico, and bottles of Sauza Hornitos Tequila were passed around with great frequency. The conversations became livelier as the evening wore on, and a great round of joke telling ensued. Everyone seemed to join in the celebration except the baptizee, who went home with her mother.

My friend had informed me earlier that the dinner would be *carne asada*. I was looking forward to that. Carne asada (broiled meat), is a popular delicacy in Mexico, especially in the northern states, where cattle raising is a major industry. Steak is cooked by broiling on either charcoal or mesquite coals. The steak is then served with *salsa,* usually the kind made with green chiles. Most carne asada is cut thinner than steaks like T-bones. Many times one will also be served *tripa de leche* (milk gut) with the steak, broiled the same way—a delicious addition.

I began to feel hunger rumbling around my stomach. I glanced at my wristwatch to find that the evening had flown and it was ten o'clock. I sought out my friend by the empty swimming pool and inquired, "Arnulfo, I don't want to sound impolite, but I am wondering when do we eat?"

"Very soon," he answered. "In fact here comes the meat for the carne asada. Look!"

I looked where he pointed and saw a cowboy, afoot, leading a yearling heifer toward the laundry room. My hopes for a quick cure for my hunger were dashed, but as I had butchered beef on many occasions, I decided to help with the procedure in hopes that I would be able to appease my hunger sooner. I must say that I have never attended a dinner party in the States where the main course was led in on the end of a rope and then butchered by the guests.

The laundry room was equipped with a drain, an accommodation for the inevitable blood from the heifer. Someone shot the animal, and the cowboy slit the jugular vein. I started the cut up the belly to remove the entrails, and as someone searched for the milk gut, I began skinning. It was not long before I had half the beef's hide skinned out, and the cowboy came over to bone out the tenderloin. One of the guests had been trying to light the charcoal and was having great difficulty. The cowboy placed the entire tenderloin on a platter beside the grill and returned to help me with the rest of the skinning operation. We were both fast with a skinning knife, and we had the carcass bare within a few more minutes. At that point I felt that I had more than contributed to the upcoming meal, so I washed my hands and returned to the party for a beer.

A roar of laughter by the swimming pool distracted me before I could reach the house. The host's two German shepherds were on the far side of the pool, making a good meal out of the tenderloin. Everyone laughed, but my hunger pangs grew. The way it happened was a mystery, except that the man trying to light the charcoal had finally succeeded, not noticing the dogs in close proximity. With the charcoal started, he went into the house for another glass of tequila, and when he returned the meat had disappeared. Another guest had noticed the dogs at the far side of the pool and commented to the charcoal tender that the dogs were really enjoying themselves. The charcoal man looked at the dogs and shrieked, "That's the meat for the carne asada!"

I went back to the laundry room, boned out the other tenderloin, and carried it out to the brazier for another try. The charcoal was perfect for broiling by this time, and the tenderloin was quickly sliced and placed on the grill along with the tripa de leche. It was a meal to remember, and not just because it was midnight when we finally ate.

Two weeks later I saw my good friend and we talked about the party.

"Did you enjoy the carne asada?" he inquired.

"I sure did," I replied. "It was absolutely delicious!"

"Do you know why it was so good?" he asked.

"I have no idea," I said. "It was just good."

"The reason the carne asada was so good," my friend continued, "is because the heifer was stolen!"

The Passing of a Friend

The day Chicaro died I lost all interest in raising horses. I still went through the motions because I didn't know that I had lost all interest. I reckon it was a matter of habit. I did what had to be done. The animals depended on me, and I wouldn't forsake that responsibility. But the enthusiasm had left me.

There is a saying: Every man deserves one good dog, one good horse, and one good woman. Chicaro was the first of that triumvirate for me, and such a joy he had been, such a good and trusting friend. I appreciated the friendship, the relationship of trust that we both held for each other. For me, it was unique.

A horsefly caused his demise. When the fly zoomed in for the attack, Chicaro rolled on the ground. It was not the bite that caused his death but his reaction. The rolling caused his intestine to twist. He was in pain, and there was no remedy. His old tissues tore when the veterinarian tried to straighten the intestine. I had to tell Doc Pickrell that it was all right to put the old stud down. I had to watch and say my farewells to the end. He is buried in his paddock. I stopped by his grave every day as long as I lived at the San Luis Ranch. I haven't returned to that spot since I left, because its memory is still vivid, but more important I can still feel the closeness I had with Parker's Chicaro. I can still hear him nicker at my approach whether or not it was feeding time. I can still see him courting his mares. I can still see him leave his stall to come and greet me. He was a gallant gentleman and will always be a part of me.

The following May I had everything going smoothly at the ranch and decided it was time to carry out my obligation to carry my father's ashes east for burial. I planned to combine the trip with an idea I hoped to propose to a major pharmaceutical company with headquarters in New York. I had designed a multiple vaccination syringe based on the same principle as a caulking gun. With this device it would be possible to vaccinate a multitude of animals without reloading the syringe. I had made rudimentary drawing of my idea and hoped to sell it.

I drove my pickup truck to the Tucson International Airport and parked it in the long-time lot because I had no idea when I would return. My ticket was one-way. The cemetery in Bourne, Massachusetts, was my first destination. I left the ashes in the care of the man in charge. With that mission accomplished, I took a train from Boston's South Station to New York City. In those days, one could reach the Biltmore Hotel from Grand Central Station, so I registered for one of their cheap rooms, one in which the communal bathroom was located at the end of the hall. The room cost eight dollars a night.

I had once met the vice-president of the pharmaceutical company in Tucson. I telephoned him from the hotel. His secretary informed me that he couldn't see me for four days. I made an appointment. Somehow, I felt I could spend four days in Manhattan. I also wanted to go to the village of Larchmont, a short train ride from the city where I had spent my first twelve years.

Every evening in Manhattan I went to a different ethnic restaurant. During the days I walked around the heart of Manhattan or spent time reading in the New York Public Library. On the morning of my appointment, I arrived at the Park Avenue skyscraper housing the company's offices a half hour early to make sure I would be punctual.

The vice-president was cordial and invited me to sit down. He had no idea why I was there, so I proceeded to explain my invention. I showed him the drawings and told him that I thought there would be a good market for such a device, especially where large numbers of animals were being vaccinated. He seemed interested. Then I made my pitch. I told him I was interested in either selling the idea to his company or becoming involved in its development as an employee. "Just a minute," he said, rising from his chair. "I want to show this to another man."

He left the office, and I sat there for fifteen minutes. When he returned, he pushed my drawings across the desk to me. "I'm sorry," he said. "My company is not interested in your idea."

With that terse statement he rose and offered his hand. "Thank you for stopping by. It was nice to see you again."

So this was the big city. So this was Corporate America. It certainly was short and not so sweet. Four days of waiting for that kind of cold shoulder boggled my mind.

Almost a year later I was reading the *Western Livestock Journal* and saw a full-page advertisement showing the same pharmaceutical company's new multiple vaccination syringe. I quickly put the pieces together. The vice-president had not conferred with anyone when he left me sitting in his office. He must have made copies of my drawings to submit to his engineering department. He probably received a substantial raise in salary. Maybe he was the biggest coyote of them all. Good old American free enterprise strikes again.

I had another mission in New York. I knew about the possibility of a manager's position on the largest ranch in Mexico, and I had the name of the man to contact in his office on Wall Street. He sounded friendly on the telephone and invited me to his office the following day, a far cry from the vice-president of the pharmaceutical company. I felt encouraged.

I took the subway to Wall Street, followed his directions, and went to his office. It was no ordinary office, at least none that I had ever seen. There were two large rooms, the smaller occupied by his secretary, a nice lady who took me in to the man I was to chat with for most of the morning. Three ticker tapes rattled on the rich maroon carpet. As he talked with me, he occasionally walked to one machine or another, picked up the tape for a glance, and returned it to its place on the floor.

The man turned out to be the uncle of the three young women who owned Rancho Santa Bárbara in Mexico, hundreds of thousands of acres of timber in Canada, and probably millions of dollars in other blue-chip assets. This was old, old money. He telephoned the niece who lived in the city, told her I was in his office, and arranged for us to have cocktails with him and attend the ballet afterward. I had a moment of pure panic. I had never met this girl and had never attended a ballet. He gave me her address, a Park Avenue apartment building, and told me, "Just give the doorman your name. He will be informed of your arrival."

The most interesting part of the conversation involved how the three nieces became owners of Rancho Santa Bárbara. When Lázaro Cárdenas was elected president of Mexico, Jack, the father of the three girls, owned a small hacienda outside Mexico City where Cárdenas and he often played polo. Cárdenas received a number of fine polo ponies as gifts from him over the years, and the two became good friends. One day when both were chatting in the presidential office, Cárdenas said that he understood that Jack was interested in buying a cattle ranch in Mexico. Jack confirmed his interest in such an acquisition. Thereupon Cárdenas went to the large map of Mexico behind his desk and with a pencil outlined an area 300 miles square in the Sierra Madres in the state of Durango. He told Jack that he should buy the property for eighty thousand U.S. dollars and not question its worth.

Six months after the purchase, Cárdenas announced that a major highway would be constructed between Durango and Mazatlán. The highway was engineered to pass next to Rancho Santa Bárbara. Of course, Jack was delighted. He set out to build a headquarters and stock the ranch with top-quality commercial cattle. He discovered that most of the cowboys in the area were afraid of the wolves that lived in the high country. The miners held no such fear, so Jack made cowboys out of miners—at least, that's how the story goes.

Jack was more than seven feet tall, and his wife was well over six feet. The daughter I met in New York equalled my own six feet three. Building the headquarters was only a small challenge for Jack. He hired stonemasons not only to construct the buildings from the native limestone but also to quarry and cut the stone into blocks. When the headquarters was finished, there were houses for the cowboys, a store, a school, and a clinic. The ranch was a modern hacienda on which some of the people had lived for twenty years without ever going outside the boundaries of the ranch.

I didn't say much during cocktails in Ted's apartment. I was glad there were other guests to carry on the conversation. I found the ballet interesting, but I must say I could have lived without experiencing it. After the ballet we were taken to the El Morocco supper club for scrambled eggs, chicken livers, and dancing. I did enjoy the dancing, especially with a good-looking girl my height. The following evening we went alone to Greenwich Village for a supper of pork chops, and we were able to talk about managing the Rancho Santa Bárbara.

I told her that I would drive to the ranch the following summer and hoped to see her there. I returned to the Biltmore and checked out. I walked over to the airline ticket building near Grand Central Station and sat down to think. A college-age girl sat next to me, and we fell into conversation. She was heading for Belgium to visit her parents, who were with the State Department. We went to the bar for a drink.

"For some reason, I don't want to go back to Nogales right now," I said.

"Do you have a passport?"

"Yes."

"How about money?" she asked.

"Plenty." I had $1,500 in cash in my pocket.

"Why don't you buy a ticket on the next flight leaving New

York for wherever in the world it might be going," she suggested as we sipped our drinks.

What a great idea! I asked her to wait at the table while I investigated. The next flight leaving New York was Aeronaves de México for Mexico City. I bought a one-way ticket, and the ticket agent filled out a tourist visa for me. I returned to the bar with my ticket, bought another drink for both of us, and for a few minutes I just sat there, in wonderment at what I had just done.

I boarded the turboprop aircraft at La Guardia and took my seat toward the front of the cabin. There were very few passengers, so I had a three-place seat all to myself. I had just fastened the seat belt when the stewardess, an attractive Mexican, asked if I would care for a drink. I ordered a Scotch. I thought it odd to be served refreshments before takeoff but concluded that perhaps Mexican airlines operated in a different manner. I finished the drink and wondered when we would be departing.

Shortly the pretty stewardess returned with another drink and the announcement that mechanics were replacing one of the generators but that the aircraft would be ready for departure in a few minutes. Having dealt with time in Mexico before, I settled back with my Scotch and waited. Finally, after a third drink, a voice came over the intercom, saying that the generator problem had been remedied and we were ready for the flight to Mexico City.

The seats in the turboprop had arms that were retractable, so when the aircraft reached cruising altitude, I flipped up the arms, curled up on the seat, and knew nothing more until the stewardess shook me to tell we were approaching Mexico City and to fasten my seat belt.

I knew nothing about the city. As far as I was concerned, it was just another city, so I told the taxi driver to take me to a hotel. The San Francisco Hotel is just off the main boulevard. I registered and went directly to my room and to bed to finish my sleep.

The dining room was still serving breakfast when I awoke at

eleven. After eating, I walked around, looking at the city and the people. But I was experiencing solitude—solitude in a morass of humanity that was in a hurry. It was the same as Manhattan, but in Manhattan I didn't feel the freedom that I did in Mexico City. Nobody knew where I was, and no matter what happened at the San Luis Ranch, Nogales, Tucson, or anywhere else, I would not know about it. And nobody could find me to tell me.

Here I had the opportunity for free thought in this unique solitude. It was unique for me because I had always thought of myself as responsible and dependable. While wandering through Mexico City, I was free from being responsible and dependable. I took advantage of those deep feelings to give thought, free thought, to where I was going, what I was doing, and what direction I might take to make a living without relinquishing my freedom.

I had tried to make a marriage work that was destined to fail at its outset because of the backgrounds of the two people involved. I felt that there had to be something done about that. I still had a strong desire for a home, but I put that aside as I ambled through the city, probably causing pedestrian jams. Then the freedom feeling, the feeling of being free, took over again. Since early youth I had wanted to be a cowboy. So I had headed out West and worked at it in Alberta, Canada, Wyoming, and Arizona. Then I had bought my own cow herd and nursed it through the drought. The Mexican steer-buying business was short-lived, as was the cotton farming operation. Now I had lost interest in raising quarter horses.

I was engaged in something I had once thought I liked, but what change could I make and remain free? I sat on a bench in a large park. I felt better out of the way of hurrying people. I had a ranch purchase deal in escrow, but somehow I knew that it would all take care of itself.

I wanted to remain responsible and dependable, but how could I maintain this present feeling and still be responsible and dependable? My thoughts turned to the possibility of managing the Rancho Santa Bárbara. I didn't think I could be free in that situation but concluded that later I would drive down to the place as I had said I would. Responsible.

Then it suddenly came to me that I didn't have to think about what to do or where to go. I would quit and let *myself* take over. Put it all behind me and see, with this new perspective, what I would think about once it was all behind, over, done with. What exhilaration I was feeling! I rose from the bench, walked to the sidewalk, and re-entered the urban scramble to walk back to the hotel.

That evening I took a taxi to the Niño Perdido district, famous for its mariachi music. All the time I had a wondrous feeling of freedom from everyone. Nobody except the young lady bound for Belgium knew where I was, and she didn't know my name.

The following morning I returned to the airport to fly to Guadalajara. I had never been to that city and wanted to see what was there. Again I asked the driver to recommend a hotel. He took me to the Hotel Fenix. I spent several days in the city, exploring and looking at the architecture, especially the huge, hand-carved wooden doors. Each afternoon I took a bus to Tlaquepaque to watch crafts being made. I sat for three hours watching a weaver work his loom. The wonderful feeling of freedom persisted. I was somewhat reluctant to leave, but I thought about my responsibilities at the ranch. I was finally ready to say to hell with all of it.

The milk-run DC-3 landed in Nogales, Sonora, and I took a taxi across the border to Escalada's, where I waited until someone came by who would take me to the Tucson airport so that I could retrieve my pickup truck and return to the San Luis Ranch. I arrived in late afternoon. The freedom feeling persisted even as I parked the pickup in the barnyard. I felt different. It was just a matter of time until I would drive out of the barnyard to see what the world might be like. I had no idea what I would do or where I might go. I did feel the exhilaration of knowing that my life had changed directions.

Guessing Weights

orman Hale and his family ranched in the mountain and hill country around Harshaw, an abandoned mining town near Patagonia. I became good friends with the Hales and helped them on roundup in the fall of 1962. Norman had shown me his neighbor Laura Bergier's ranch, which was for sale, and I had made an offer on it. As I mentioned earlier, the transaction was in escrow, and I had looked forward to becoming Norman's neighbor.

I didn't want the cattle that were on the ranch, for two reasons. The first was that if I bought the cows, I would have to pay cash. Cattle are always cash, and I didn't want to sell any securities to raise the money. The second reason was that I didn't like the animals anyway; there were too many old cows. Beyond this, the range looked overgrazed. I wanted to rest it and let it re-seed itself through a couple of rainy seasons before restocking the range. Overall, however, I felt the price I was paying was reasonable. I had a plan in mind for the future that involved weaning the calves on the Bergier ranch, and bringing them to the irrigated pastures on the San Luis. The plan coincided with the two-year rest for the Bergier place in that I figured it would take those two years to have the San Luis Ranch fully developed into irrigated pastures.

I consulted with the forest ranger in charge of grazing permits to see if I could still maintain the permit on Coronado National Forest land while resting the ranch. The conversation seemed vague on his part when it focused on my applying for a "non-use" permit. I decided I would pay the regular grazing fee rather than risk getting further involved in government bureaucracy.

I helped with the roundup after the herd was sold to a man in Arivaca. I was happy to see that happen because he had paid more for the cattle than I would have, and the widowed owner needed all the resources she could gather.

With that roundup over, I helped Norman gather his herd and ship his calves. We began early in the morning with plenty of help, so by noon we had all of his cattle in a holding trap, ready to separate calves from cows and put them on trucks to weigh them in Patagonia. Before going to the Hale headquarters for lunch, I rode slowly through the cattle in the holding trap, looking them over and estimating the weights of his calves.

Norman's wife, Ruth, had fixed a delicious meal of *carne con chile,* which we ate with relish. We were sitting at the table afterwards, sipping our coffee, when Norman expressed worry that the calves would be light in weight.

"I'll bet the calves won't average two eighty-five," he said.

"Norm, you're way off," I said. "I rode through 'em, and I'll bet you a six-pack of Michelob that those calves will average three hundred and sixty-five at the scales in Patagonia."

"I'll bet you the six-pack, but I hope you win it," he said.

We returned to the holding trap, drove the cattle into the corrals, cut the calves away from their mothers, and drove them into the trucks.

I didn't stand by the scales as the calves were weighed because I was working the corrals. When the last bunch left the scales, Norman added up his figures and calculated the average weight of the animals. His eyes brightened and a broad smile beamed from his face.

"You really have a good eye, John," he said. "They averaged three sixty."

"That's great!" I replied. "I'll even buy the six-pack after that."

I was really glad that I had been so close to the average weight because Norman needed all the pounds he could get from his calves.

About two weeks after I returned from Mexico City, Norman told me that someone else was interested in buying the neighboring ranch and would be willing to pay $10,000 more than my contract of sale called for. I knew my contract would hold as long as I came up with the down payment, and that was sitting in a savings account, ready.

I had met with my attorney in Tucson to go over the legal part of the transaction. He seemed like a very happy man with a nice wife and several children.

"You make more money having a good time than I'll ever hope to earn practicing law," he said during our conference.

"I'll tell you what," I replied. "I would trade you straight across anytime. You have a home and family. I may make money having a good time, as you put it, but I would much rather have a home."

With the thoughts I had pondered on the streets of Mexico City, I decided to go to Laura Bergier's lawyer in Nogales and tell him that if the other couple wanted to buy the ranch, I would step aside as long as my earnest money was returned. He agreed to my offer.

One night I returned from town just after midnight to find Little Burr, the last mare to be bred to Chicaro, down in the paddock in labor. I kept the headlights of my truck trained on her and went in to see how she was proceeding. One of the foal's hooves had started out, and I thought I saw its tongue. It looked like a normal birth except that the other hoof had not made its appearance. The mare got to her feet, circled once, and went down again, thankfully with her rear end toward the headlights. I sat down behind her and reached inside to find the second hoof. It was there, but she was having trouble getting the foal's shoulders through the birth canal.

I waited for her labor to subside, grabbed both of the foal's legs, and pushed in with all the strength I could muster. When I had the foal as far back as I was going to get it, I pushed on one leg and pulled on the other, outward and downward, trying to get the shoulders through. I knew if that could be accomplished, the foal would slide out in a second. When the mare started into labor again, I braced my feet against her gaskins, just above the hocks, and pulled again. The mare pushed and I pulled. All of a sudden the foal came sliding out, and I fell flat on my back with the little filly on top of my stomach. I looked straight into her mouth, soaked to the skin with amniotic fluid, but I didn't care about anything except feeling the joy of having a live foal on top of me.

I left the pair to do what they had to do while I went to the house for a hot shower. I returned an hour or so later to find the new-born Chicaro filly nuzzling her mama's bag. Little Burr nickered as I approached. "Good job, gal," I said. "You did old Chicaro proud."

I wondered again about becoming manager of Rancho Santa Bárbara. The trip to Mexico had begun in June. I remember writing in my journal, "Hermosillo is burning." I spent one night in the San Alberto Hotel in Hermosillo before driving south to Navojoa to meet Joe Brown. I called him from the Motel del Río, located near the south bank of the Río Mayo.

The first time I had met Joe was in the El Dorado bar in Nogales, Arizona. We had had some good times over the several years since, and we were friends. Joe drove over to the Motel del Río, and we had a great evening of conversation and Scotch whiskey. I followed him back to his house and slept on the couch in his living room. In the morning I looked up at a large framed diploma over the fireplace that read, "Bachelor of Arts in Journalism, Notre Dame University, Joe Paul Summers Brown." This man was not just buying corriente cattle in the Sierra Madres, he was gathering what was to become the most unique sense of the country and the people that has ever been written down.

His first book, *Jim Kane,* became a motion picture. The book is a masterpiece, but like many movies, *Pocket Money* fell short, very short, of capturing the essence of Joe's writing.

My trip involved two different projects. First, I tried to determine a location for a feed-mill operation on the west coast of Mexico around either Navojoa, Los Mochis, or Culiacán. After spending two weeks investigating, I concluded that for a gringo to attempt such an enterprise would be risky at best. The second part of the trip involved Rancho Santa Bárbara.

From Navojoa I traveled to Mazatlán, and after a brief sojourn in that resort city I left for the top of the Sierra Madres. The highway heads east from Reunión, passes through the furniture town of Concordia, and then swirls its sinuous way up the mountain. The narrow, boulder-strewn highway made for slow, cautious driving. About halfway I saw two workmen gathering the boulders and throwing them into a highway department truck. I suddenly realized that the boulders littering the pavement had rolled off the steep mountain slopes during a heavy rain the previous night. What if I had decided to travel the night before instead of waiting until morning?

The hamlet of El Salto sits at the top of the highway before it levels off to pass through the Rancho Santa Bárbara. I had driven there for two reasons. The first was to look at the property as a potential investment for the same man who wanted to start a feed-mill operation on the west coast of Mexico. The second reason for my visit was to decide if I wanted to become the manager.

I waited at the entrance for a man to walk from his house to unlock the large iron gate. Tending the entrance was the man's only responsibility. I drove for a few miles to the headquarters, where I was met by the manager, who was supposedly planning to retire. He showed me through the buildings of native limestone with doorways ten feet tall and ceilings that ranged from fifteen to twenty feet. At supper I discovered that he had no intention of retiring and had recently hired a young American to help him with the day-to-day overseeing. I wondered if the owners were planning to force the man to retire after twenty years in their employ. I felt that I had driven those many miles for nothing.

I spent a week driving around with the manager and still saw only part of the ranch. He told me that the girls were asking three million U.S. dollars for it. I asked myself, why had I been told that they wanted me to look at the ranch with the idea of managing it? When I left I decided to drive into Durango to see what I could find out about Rancho Santa Bárbara from someone who had no financial interest in it.

Like the El Dorado in Nogales, Arizona, the Hotel Casa Blanca was the place where cattlemen congregated. I took a room and spent a day and two nights getting acquainted. I asked a few innocuous questions and noted that several men mentioned rumors that the *ejiditarios,* the land reform group, had their sights on the ranch for an invasion and subsequent division into ranching ejidos.

That was enough for me. The entire business confused me completely. There was the nebulous job as manager and then the threat of an ejiditario invasion. I wanted no part in dealing with Mexican land reformers. I drove back to Tucson through Chihuahua and out through El Paso. I saw a lot of country I probably never would have seen otherwise, but I couldn't figure out who the coyote was.

I sent a "thanks but no thanks" letter to New York. I also advised my client that purchasing the ranch could mean financial disaster if the place became a community of ejidos.

Then I went over to see Joe Escalada at his store in Nogales. We talked for a while, as usual. I told him about the freedom I had felt since my trip to Mexico City. "With that feeling inside of you," he said, "you can do anything you set your mind to."

In Retrospect

To many a two-legged coyote, business is a game, especially when his or her survival is not at stake in a particular deal. And, of course we must not forget what seems to be a prevalent human trait: greed.

We must also distinguish between a coyote and an out-and-out confidence artist, who deliberately calculates an outright swindle. I would not think of using the term *coyote* to describe these people. I asked several people if they could distinguish between a coyote and a con artist, and the best answer was that a con artist deliberately tries to enrich himself or herself by swindling and lying, whereas a coyote is just trying to make a living. That may seem to let the coyote off the hook too easily, but in light of what I have experienced, it may be the best way to describe one.

I look back at the coyotes who have crossed my path with a certain cynicism that I don't feel so good about because it is not my desire to be a cynic. When I deal with coyotes, I take them in stride and at times mimic them for my own survival. These are the times I find challenging. So are all of us coyotes? Or, at least, at times do we sometimes take on a coyote's character traits? Certainly some are more coyote-like than others. Perhaps we must look at the canine coyote as a symbol of survival in a world dominated by competition. I have tried to eliminate competition from my life, but that may be far easier than eliminating competition from my psyche.

All the enterprises that I attempted during the years I have described in this book were hampered by my lack of substantial capital. I recognized that I needed more volume to succeed in the Mexican steer-buying business, but that would have taken more capital than I was willing or able to come up with. Order buying is highly competitive, and I had neither the broad experience that would have given me enough contacts nor the inclination to haggle over prices.

The system of buying Mexican cattle for export to the United States has changed considerably since the time I wandered around northern Sonora buying steers. Then the risks were many, from the arbitrary closing of the border to cattle exports by the Mexican government to untimely price fluctuations. Somewhere in the midst of all this, the coyotes always seemed to lurk.

Transporting livestock from distant areas was primarily done by rail until 1972, when a man named Aguirre went into business with the first double-decker cattle truck. Today, few if any herds are driven to the crossing points by men on horseback, because the steers are bought on the basis of duty weight rather than by the head. The cattle being exported are always weighed, because both Mexican and U.S. customs duties are based on weight.

Communications have also improved considerably. During the early 1960s, the village of Tubutama in the Altar Valley had but one telephone, as did Sásabe. To contact a rancher in those days, a person had to travel to the ranch, mostly over very primitive roads, or find the rancher, by chance, in town. Highway travel is far easier today with the four-lane, divided toll roads, or at least paved highways. The paving of the road from Sásabe to Altar is planned, and there already is a paved road from Altar to Saric. These improvements aid livestock transportation and make for easier access to rural Sonora. The same is true in the United States, with its interstate highways.

There is still a demand for Mexican cattle in the United States. The fact that Mexican steers make good weight gains when put in a different environment makes them attractive to many U.S. cattlemen and feedlot operators. However, the U.S. buyers rarely buy the cattle in Mexico, instead relying on one exporter in Nogales for 90 percent of their steers. They would rather pay at least fifty cents a hundredweight over the duty weights and take delivery in the United States than face all the risks of doing business directly with Mexican ranchers.

Life along the border is far different today from what it was during the early 1960s. The drug traffic from Sonora into the United States has taken its toll on a former way of life that was less dangerous to the steer buyer. The old camaraderie between American buyers and Mexican ranchers seems to have been diluted. The places we once frequented to meet and do business are no longer filled with men in broad-brimmed hats. More evident are people engaged in the produce business or connected to the maquiladoras.

The maquiladoras are responsible for a great migration of Mexicans from farther south to the border cities like Nogales, Sonora, in search of employment in these plants. Many migrate with only a hope of finding work, and many end up living in the squalor of shanty settlements on the outskirts of the cities even if they do find work. These hovels are made from every sort of scrap material available. The more elaborate may incorporate pieces of used plywood, but many are made of cardboard. Sanitation is nonexistent.

With this swelling population comes an increase in crime on both sides of the border, and environmental stress such as excess sewage that ends up in Arizona. The border culture is unique to both countries, but the recent migration of Mexicans to work in the maquiladoras and the migration of Americans pouring in to manage those industrial plants, participate in the produce business, or staff the Border Patrol and the customs offices have all contributed to the changes. The unique quality of the culture remains; it has just changed with time.

The instability of the Mexican currency raises hell with most commerce on the U.S. side of the border in cities like Nogales, Arizona. When the peso is devalued, Mexican patrons of Nogales stores can no longer afford to shop in Arizona. Although the cattle trade is always dependent on the price of beef in the United States in U.S. dollars, it is government regulations and actions that can really wreak havoc with the cattle business. When the Mexican government closes the border to cattle exportation, "because there is a shortage of beef in Mexico City," one must suspect there is another reason for the closure. The closures create hardship for the cattlemen in northern Mexico but do little or nothing to influence the supply of beef in far away Mexico City.

It is difficult to predict how NAFTA, the North American Free Trade Agreement, will affect the cattle trade between the United States and Mexico. Many of my long-time friends along the border seem to feel that the only segments of both populations that will benefit from the agreement are the wealthy. The expression, "What does NAFTA stand for? North Americans Forgot The Alamo," is popular among many who have doubts about the agreement.

The current trend in the United States of ranchers battling against government agencies that control grazing on the public lands and environmentalists who feel that all cattlemen are ruining the land may have far-reaching effects on the cattle trade across the border. If the government and environmental groups succeed in forcing western U.S. cattlemen out of business, there will be a greater demand for Mexican beef in the United States.

Since the 1960s, a government-sponsored feedlot in Carbó and one in Hermosillo have been established. The Mexican government has also instituted a beef grading system after the U.S. model. Mexican butchers now follow the cuts of beef long used in the United States as opposed to the old way of "boning out" a carcass and slicing chunks of meat with the grain instead of across. Meatpacking plants, such as the one that sprang up in Magdalena during the Aftosa quarantine, may one day become popular again with an eye toward the U.S. market.

I learned a lot by doing what I did, when I did, and with whom I did it. And as is true with the rest of my life, I have no regrets. In fact, these stories are fond memories I am happy to share. The fight for survival seems to be a common instinct among species. Some are more successful than others.

Index

Aeronaves de México, 132
Agua Nueva, Rancho, 4
Alberta (Can.), 11
alfalfa, 34, 35, 39, 40, 92
Altar, 16, 18, 19, 142
Altar River, 17, 19
Arivaca Land and Cattle Company, 112
Arizona Livestock Sanitary Board, 112
Arizona ranch, La, 112
Arizona State Fair Quarter Horse Show, 72
Atil, 19
auctions, livestock, 59, 109–10

beef, canned, 27
Beula Burns, 72
Bogan and Bernard, 112
Bond, Paul, 111
border, U.S.–Mexico, xi, xii, 5–7, 15, 38, 103, 104
   closing of, 26–27, 29, 31
   crossing livestock over, 39–40, 42–45, 84
   life on, 77, 123, 134, 142–44
   overgrazing at, 17, 91
   ranches on, 13, 63

Border Patrol, 95–96
Bourne (Mass.), 128
Brahman cattle, 6, 112
Brangus cattle, 56, 110
Brown, J.P.S. (Joe), 138
Buenos Aires Ranch, 13

Cabezón. See Wooddell, Les
Caborca, 18, 19, 90
Camino del Muerte, 12
Canoa Ranch, 39, 54–55, 59–63
Carbó ranch, 112
Cárdenas, Lázaro, 130
careless weed, 31, 32
Castillo, Luis, 92–94
Castillo, Rafael, 4, 26, 30, 39, 63
Castillo Ranch, 94
cattle, 57, 87
   border crossings with, 3–4, 32–33, 35, 38–42, 43–44, 62, 91
   breeds of, 5, 20, 56, 97, 110, 138
   crossbred, 6–7, 22–23
   dipping, 42–43, 45
   feedlots and, 60–61, 63, 91, 97
   grazing permits for, 135–36

cattle (*continued*)
    managing, 58, 79, 88–89, 112
    raising, xi, 50, 78, 92, 111,
       123, 135
    trading, 74, 83, 96–98, 105,
       144
    transporting, 142–43, 144
    value of, 136–37
cattlemen, 112–13, 140, 143–44
Cattlemen's Club (Phoenix),
    84–85
Charolais cattle, 112
Chicaro. *See* Parker's Chicaro
Chicaro Bill, 72
Christmas, 27
corn, 11, 29, 58
Coronado National Forest, 136
corrals: estacada, 21, 34, 61
corrientes, 6, 138
cotton: 10, 12, 21, 29, 54, 58, 111,
    133
    choppers, 31–32, 50–51
    picking, 51–53
coyotes, 52, 95, 140
    corporate, 128–29
    defined, xi–xii, 47
    in horse trading, 74–76,
      101–3
    learning from, 141–42
    real estate and, 69–70
Cuauhtémoc Reservoir, 19
currency, Mexican, 14, 144

Desert Homestead Act, 16
drought, 29–30, 32, 34–35
Dupré, Carlos, 83–84

ejidos, 4, 15, 17, 140
El Dorado Bar, 74–77
Elías, Francisco, 112
El Pinto, 40–41
El Salto, 139

El Tecolote, 77
Escalada, José (Joe), 111, 112
Escalada, Luis, 112
Escalada, Manuel, 112
Escalada Ranch Supply store, 111,
    112, 134
Estrella, Chema, 74–75, 76–78

farming, xi, 9, 11, 16–17, 19, 29,
    57–58
feedlots, 59–61, 63, 91, 97, 143–44
Felix Gómez ranch, 87
Fisher, Lee, 4
flying, 44–45, 89–90
foot-and-mouth disease, 26–27

Garrett, Jim, 27, 32, 62
grazing permits, 135–36

hay, 10–11
Herefords, 3, 5, 20, 22–23, 88, 97
highways, 12, 130
horses, 11
    breeding, 72–74, 120–21
    racing, 119–21
    trading, 74–76, 101–3
Hudgins Air Service, 30

imigrados, 7
Imperial Valley, 17

Januske, Fr. 18
Jarillas Cattle Company, 112
Johnson grass, 31

Kino, Eusebio Francisco, 17, 18
Knagge, Mike, 45, 113

La Arizona ranch, 112
La Matanza, 19
land reform, Mexican, 15, 140
La Osa Ranch, 42–44, 63

La Reforma, 18
Larchmont (N.Y.), 128

McVey, Jack, 112
Magdalena, 27, 32–33, 46, 145
Manhattan, 128, 129, 130, 131, 132
maquiladoras, 143
Marshall Plan, 27
Matanza, La, 19
Mazatlán, 130
Mexicali, 88
Mexican Revolution, 15
Mexico, 6, 14, 33, 95, 113, 114
    cattle trade in, xi, 3–4, 5–7,
        10, 63, 96, 119, 133
    customs in, 27–28, 38, 41,
        104
    government regulations in,
        26, 32, 104, 142, 144
Mexico City, 26, 130, 132–33, 140,
    144
migration, to border, 143–44
Mrs. Thomison II, 73–74
Montez, María, 10, 11
Montez, Rudolfo, 10, 11
mordida, 7, 104
morning glory, 31
Mosco Bampo Ranch, 114

Newcomer, Floyd, 55–56, 64
North American Free Trade
    Agreement, 144
Nogales (Ariz.), 87, 99, 103, 104,
    105, 111, 123
    El Dorado Bar in, 74–77
Nogales (Son.), xii, 32, 33, 37, 77,
    103, 104
Nogales International Airport, 87

Oacpicagigua, Luis, 17
O Bar J herd, 110, 111
Opata, 18

Oquitoa, 19
Ortiz Padilla, Diego, 17
Osa Ranch, La, 42–44, 63
overgrazing, 17, 91

Papagos, 112
Parker, Wirt D. "Dink," 72
Parker's Chicaro, 72–74, 128–29,
    133
    ads for, 100, 101
    buying, 75–76
Percherons, 11
Piggin String, 120
Pima Revolts, 17, 18
Pinto, El, 40–41
Pioneer Hotel, xi, 9
poverty, 15
Pozo de Crisanto Ranch, 112
Pozo Verde, 26
Producer's Cotton Gin, 51
Purina, 31

QHB Magazine, 100, 101
Quarter Horse Journal, 120–21
quarter horses: xi, 66, 119
    breeding, 72–74
    trading, 75–76, 101–3
Queeny, 120

racing, horse, 101–2, 119–21
radio, 99–100
Rebeil, Pete, 62
Reforma, La, 18
ranches. See by name
Rillito Race Track, 101–2, 119–20
Rodríguez, Toribio, 4, 5, 13, 17,
    19, 20
Rukin String, 120, 121

Sahuarita, 51–52
Salto, El, 139

San Luis Ranch, 66–68, 100, 109
    leasing, 107–8
    Parker's Chicaro on, 72–74
Santa Ana, 19
Santa Bárbara, Rancho, 130–31
Santa Cruz County Jail, 96
Santa Cruz River, 66–67, 92
Santa Cruz Valley, xii, 31–32
Santa Margarita Ranch, 114
San Vicente, Rancho, 6, 7, 17, 24, 25
    accommodations on, 20–21
    cattle on, 38–39
    drought on, 29–30, 32, 34–35
    location of, 4–5
Saric, 4, 13, 16–18, 20, 22–24, 142
Saric, Luis de, 17
Sasabe (Ariz.), 5, 25, 45, 46, 114
Sásabe (Son.), 4–5, 24, 27, 30, 62–63
    border crossing at, 31, 40–46, 114
Sasabe road, 44
Sasabe Store, 41
Saxon, Harry, 120
Shorthorns, 6
Sierrita Mountains, xi
smuggling, 104
Social Security, 17
Sonora, Río, 18
sorghum, Hegari, 11, 29, 58
Soto, Miguel, 16–17
Soto, Sara, 16, 17
Southern Arizona Livestock Association, 109–10
Southwest Futurity, 119
steers. See cattle
Sykes, Gene, 112

Tecolote, El, 77
tick control, 42–43
trade, international, 103–4, 143–44
Tuape, 18
Tubutama, 18–19, 142
Tucson, 6, 30, 59, 67–68, 72, 105, 140
    commuting from, 11–12
    horse racing in, 101–2, 119–20
    marketing, 61–63
Tucson International Airport, 30, 128
Tucson Mountains, 116
Tucson–Nogales Highway, 12

United States
    cattle importation into, 3, 7–8
    international trade and, 103–4
    Mexican border and, xi, 26, 27, 43–45, 123, 141–44

Villa, Pancho, 15

White House department store, 51
Willcox, 120
Wooddell, Les (Cabezón), 112–14
Wyoming, 10

Yaqui Indian Reserve, 112
Yaquis, 113–14
Yuma, 55, 64

Zapata, Emiliano, 15
Zúñiga, Roberto (Bobby), 83, 84–87

About the Author

John Duncklee has pursued a varied life, working as a cowboy, sailor, university professor, mesquite furniture designer, and writer. He graduated from the University of Arizona in 1956 after a four-year hitch in the navy during the Korean War. After a number of years spent in cattle ranching, buying steers in Sonora, Mexico, farming near Tumacacori, consulting on the ranching business in Mexico, and breeding quarter horses on a ranch near Nogales, Arizona, he returned to the University of Arizona to attend graduate school in geography. He has taught at the University of Arizona, Northern Arizona University, and the Universidad de Sonora in Hermosillo.

While living in Flagstaff, Duncklee wrote the environmental impact statement *Man—Land Relationships on the San Francisco Peaks,* which was influential in halting an attempt to develop 160 acres of Hart Prairie on the western slope of the peaks. In conjunction with this effort, he wrote the lyrics and some of the music for the phonograph album *Did You Ever Sing to a Mountain.*

Since 1973 Duncklee has been a freelance writer and has had articles published in *Arizona Highways,* the *Christian Science Monitor, Defenders of Wildlife,* the *Journal of Irreproducible Results,* and the *Tombstone Epitaph.* His monthly column, "View from the Porch," appears in *Connection,* which is published in Arivaca, Arizona.

www.ingramcontent.com/pod-product-compliance
Lightning Source LLC
Chambersburg PA
CBHW020432290526

45785CB00002B/815